SpringerBriefs in Applied Sciences and Technology

Computational Intelligence

Series Editor

Janusz Kacprzyk, Systems Research Institute, Polish Academy of Sciences,
Warsaw, Poland

SpringerBriefs in Computational Intelligence are a series of slim high-quality publications encompassing the entire spectrum of Computational Intelligence. Featuring compact volumes of 50 to 125 pages (approximately 20,000-45,000 words), Briefs are shorter than a conventional book but longer than a journal article. Thus Briefs serve as timely, concise tools for students, researchers, and professionals.

Khalid Raza

Machine Learning in Single-Cell RNA-seq Data Analysis

 Springer

Khalid Raza
Department of Computer Science
Jamia Millia Islamia
New Delhi, Delhi, India

ISSN 2191-530X   ISSN 2191-5318 (electronic)
SpringerBriefs in Applied Sciences and Technology
ISSN 2625-3704   ISSN 2625-3712 (electronic)
SpringerBriefs in Computational Intelligence
ISBN 978-981-97-6702-1   ISBN 978-981-97-6703-8 (eBook)
https://doi.org/10.1007/978-981-97-6703-8

To my Family & Friends
For their unwavering love, encouragement, and understanding during the countless hours I spent researching, writing, and pondering the complexities of both machine learning and single cell.

To Researchers & Scientists
To all the researchers, data scientists, and bioinformaticians who are dedicated to exploring the complexities of genomics through the lens of machine learning. May this book serve as a valuable resource in your efforts to decode the vast landscape of single-cell RNA sequencing data.

To My Students
To my students, past and present, who have challenged me to think critically and creatively. Your curiosity and passion for learning are what makes teaching such a rewarding endeavor.

This book is dedicated to all the readers of this book, whether you are a seasoned professional or a curious learner. May this work ignite your passion for discovery and empower you to contribute to the ever-evolving field of Bioinformatics & Computational Biology.

With heartfelt gratitude,

Khalid Raza

Preface

Single-cell RNA sequencing (scRNA-seq) represents a transformative technology that allows us to profile gene expression at the resolution of individual cells, providing unprecedented insights into cellular heterogeneity and complex biological processes. By capturing the transcriptomic landscape of thousands to millions of individual cells, scRNA-seq has revolutionized our understanding of developmental biology, disease mechanisms, and therapeutic responses. However, the high-dimensional, sparse, and noisy nature of scRNA-seq data presents unique challenges for its analysis and interpretation. Traditional computational methods often fall short in dealing with the complexity and scale of these datasets. This is where machine learning, with its powerful data-driven approaches, comes into play. Machine learning algorithms are adept at uncovering patterns, making predictions, and inferring biological insights from large and complex datasets.

In this book, we aim to provide a comprehensive overview of the application of machine learning techniques in the analysis of scRNA-seq data. We explore how these advanced computational methods can enhance our ability to identify cell types and states, analyze differential gene expression, and infer cell trajectories. The contents of this book are structured to guide the reader through the essential concepts and methods used in the field. We begin with an introduction to single-cell RNA sequencing technology and the fundamental challenges associated with its data analysis. Subsequent chapters cover data preprocessing, quality control, dimensionality reduction approaches, and other application-specific tasks. Various machine learning approaches, including unsupervised learning for clustering and visualization, supervised learning for classification and prediction, and deep learning techniques for more complex inference tasks have been discussed.

Specific topics include:

Data Preprocessing and Quality Control: Techniques for filtering, normalization, and dimensionality reduction to prepare scRNA-seq data for analysis.

Clustering and Cell Type Identification: Machine learning methods for identifying distinct cell populations and characterizing their unique gene expression profiles.

Differential Expression Analysis: Basic concepts of differential expression analysis, its methods, tools, and software, along with case studies have been presented.

Trajectory Inference and Cell Fate Prediction: Algorithms for reconstructing developmental lineages and understanding dynamic processes within single-cell datasets.

Emerging Topics and Future Directions: Explores cutting-edge topics and future directions in single-cell RNA-seq data analysis also discusses the challenges and opportunities that lie ahead, encouraging readers to engage with the ongoing evolution of the field.

The purpose of this book is to equip researchers, data scientists, and students with the knowledge and tools necessary to harness the power of machine learning in scRNA-seq data analysis. By presenting state-of-the-art methods and their applications, we aim to foster a deeper understanding of cellular heterogeneity and contribute to the advancement of personalized medicine and targeted therapies. I hope this book serves as a valuable resource for those engaged in the fascinating and rapidly evolving field of single-cell genomics. May it inspire new ideas, foster innovative research, and ultimately contribute to the betterment of human health through the integration of cutting-edge computational and biological sciences.

New Delhi, India   Khalid Raza

Acknowledgments

The completion of this book on *Machine Learning in Single-Cell RNA-seq Data Analysis* would not have been possible without the support, guidance, and contributions of many individuals and institutions. I am deeply grateful to all those who have been part of this journey.

First and foremost, I extend my heartfelt thanks to my family for their unwavering love, patience, and encouragement. Your support has been my anchor, enabling me to pursue my passion and complete this work.

I am immensely grateful to the researchers and scientists whose pioneering work in single-cell RNA sequencing and machine learning has laid the foundation for this book. Your innovative contributions have inspired countless advancements in the field, including my own. I also wish to acknowledge the collaborative efforts of the broader scientific community. Their dedication to advancing knowledge through shared research and open dialogue has been a source of motivation and inspiration.

To my students, past and present, thank you for your curiosity, enthusiasm, and willingness to explore new ideas. Your questions and perspectives have challenged me to think more deeply and creatively about the topics covered in this book.

A special acknowledgment goes to our publisher Springer, editors, and production team for their dedication to bringing this book to fruition. Your efforts in editing, design, and production have been crucial in bringing this book to fruition.

Finally, I am grateful to the readers of this book. Whether a researcher, data scientist, student, or curious learner, I hope this work provides them with valuable insights and tools to advance the understanding and application of machine learning in single-cell RNA-seq data analysis.

Disclaimer I have used the large language model (LLM), ChatGPT to refine English and grammar of the contents, and summarize them. The prompt I have used is "Refine language and correct grammar for better clarity", and "summarize the contents". All the information has been validated and checked upon language refinement.

Khalid Raza

About This Book

Single-cell RNA sequencing (scRNA-seq) has revolutionized our ability to study gene expression at the resolution of individual cells, providing deep insights into cellular heterogeneity and complex biological systems. This book, *Machine Learning in Single-Cell RNA-seq Data Analysis*, offers a comprehensive guide to leveraging machine learning techniques for the analysis and interpretation of scRNA-seq data. It is designed for researchers, data scientists, and students who seek to harness the power of machine learning to unlock the full potential of single-cell genomics.

The book is organized into six meticulously crafted chapters, each addressing a critical aspect of scRNA-seq data analysis:

Chapter 1: Introduction to Single-Cell RNA-seq Data Analysis

This chapter provides a foundational overview of single-cell RNA sequencing technology. It covers the principles and methodologies of scRNA-seq, highlights its significance in modern biology, and discusses the unique challenges posed by single-cell data. Readers will gain a clear understanding of the landscape of scRNA-seq analysis and the role of machine learning in addressing its complexities.

Chapter 2: Preprocessing and Quality Control

Accurate analysis of scRNA-seq data begins with robust preprocessing and quality control. This chapter delves into the essential steps for preparing single-cell data for analysis, including filtering, normalization, and batch correction techniques. It emphasizes the importance of addressing technical variability and data sparsity to ensure reliable downstream analyses.

Chapter 3: Dimensionality Reduction and Clustering

Understanding cellular diversity requires effective dimensionality reduction and clustering methods. In this chapter, we explore various machine learning techniques for reducing the high dimensionality of scRNA-seq data and identifying distinct cell populations. Methods such as Principal Component Analysis (PCA), t-Distributed Stochastic Neighbor Embedding (t-SNE), and Uniform Manifold Approximation and Projection (UMAP) are discussed, along with clustering algorithms like k-means, and graph-based approaches.

Chapter 4: Differential Expression Analysis
Differential expression analysis is crucial for identifying genes with significant changes in expression across different cell types or conditions. This chapter covers statistical models and methods specifically designed for single-cell data, including DESeq2, edgeR, MAST, and more. It provides practical guidance on performing differential expression analysis and interpreting the biological significance of the results.

Chapter 5: Trajectory Inference and Cell Fate Prediction
Single-cell RNA-seq allows for the reconstruction of developmental trajectories and prediction of cell fates. This chapter introduces machine learning approaches for trajectory inference, such as Monocle, Slingshot, and PAGA. It discusses methods for ordering cells along developmental timelines and predicting cellular transitions, offering insights into dynamic biological processes. It also provides practical demonstration on trajectory inference using PAGA.

Chapter 6: Emerging Topics and Future Directions
The final chapter explores cutting-edge topics and future directions in scRNA-seq data analysis. It highlights recent advancements in multi-omics integration, spatial transcriptomics, and the use of deep learning for more complex inference tasks. The chapter also discusses the challenges and opportunities that lie ahead, encouraging readers to engage with the ongoing evolution of the field.

By bridging the gap between single-cell genomics and machine learning, this book equips readers with the knowledge and tools necessary to tackle the intricacies of scRNA-seq data analysis. Through detailed explanations, practical examples, case studies, and Python program it aims to foster a deeper understanding of cellular heterogeneity and contribute to the advancement of personalized medicine and targeted therapies. Whether you are a novice or an experienced practitioner, this book provides a valuable resource for navigating the exciting and rapidly evolving intersection of computational biology and single-cell genomics.

Contents

About the Author

Dr. Khalid Raza is an **Associate Professor** at the Department of Computer Science and **Deputy Director** of Malaviya Mission Teacher Training Centre (Formerly known as UGC-Human Resource Development Centre), Jamia Millia Islamia (Central University), New Delhi, India. He is also an **Honorary Research Fellow** at INTI International University, Malaysia. Dr. Raza was on foreign deputation during the academic session 2017–2018 and served as an **ICCR Chair Visiting Professor** at the Faculty of Computer and Information Sciences, Ain Shams University, Cairo, Egypt. Dr. Raza has over 13 years of teaching and research experience at both undergraduate and postgraduate levels. He has **contributed over 140 peer-reviewed research articles** in refereed international journals, conference proceedings, and book chapters, having a **cumulative impact factor over 250**. Dr. Raza has authored/edited more than 10 books published by reputed publishers like Springer-Nature, Elsevier, and CRC Press. Dr. Raza is an **Academic Editor** of *PeerJ Computer Science* and **Guest Editor** of *Natural Product Communications*, and *Mini-Reviews in Medicinal Chemistry*. He has an active collaboration with the scientists from leading institutions of India and abroad. He has received grants for Government funded research projects and worked as Principal Investigator. He is also recipient of BRICS Multilateral research grants and working on the WHO's high priority area along with other collaborators from CSIR-IGIB (New Delhi), and UFMG (Brazil). Recently, Dr. Raza has received research grants from SERB (DST) and is working as Co-PI. He is recipient of the Indian Academy of Sciences Faculty Summer Research Fellowship. He is a reviewer of several international journals, a member of several conference review committees. Dr. Raza has been featured in the list of **World's Top 2% Scientists** released by Stanford University (USA) in collaboration with Elsevier for the two consecutive years 2021 and 2022. He has also delivered several invited talks in International conferences, and workshops (AICTE-ATAL, UGC-HRDC Centres, etc) and inspired thousands of students and young faculty members. His research interest includes *Computational Intelligence* methods and their applications in *Bioinformatics* and *Health informatics*.

Chapter 1
Introduction to Single-Cell RNA-seq Data Analysis

1.1 Single-Cell Sequencing

1.1.1 Overview

DNA sequencing is a method for determining the exact order (or sequence) of nucleotides, or bases, in a given DNA fragment. There are four base pairs of nucleotides, represented by the first letter of their chemical names such as A (Adenine), C (Cytosine), G (Guanine), and T (Thymine). This sequence of nucleotides encodes biological information that cells utilize for their development and growth. Similarly, RNA sequencing (RNA-seq) is a process of identifying the order (or sequence) of the bases that make up an RNA molecule. Generally, RNA-seq is used to identify the quantity of RNA in a sample, i.e., allowing us to analyze the transcriptome, and tells us which genes are expressed in a cell, and to what extent. RNA-seq helps researchers perform gene expression profiling, single nucleotide polymorphism (SNP), RNA editing and differential expression analysis. The gene expression profile represents cellular activities, mostly used to study diseases, cellular behaviour, and response to drug or other stimuli. The sequencing methods have been categories into different generations: from Sanger sequencing (first generation) to next-generation sequencing (NGS). A detailed discussion on the history and various available sequencing platforms can be found in Raza and Ahmad [42].

Single-cell sequencing (SCS) is a suite of technologies that allow researchers to sequence the genomes or transcriptomes of individual cells. This is in contrast to traditional sequencing methods, which sequence the DNA or RNA from a bulk population of cells [25]. In other words, single-cell sequencing is a cutting-edge and powerful molecular biology technique that allows researchers to analyze the genetic and genomic information of individual cells within a complex mixture [55]. This technology has revolutionized our understanding of cell biology, enabling scientists to explore the heterogeneity of cell populations, identify rare cell types, and gain

K. Raza, *Machine Learning in Single-Cell RNA-seq Data Analysis*, SpringerBriefs in Computational Intelligence, https://doi.org/10.1007/978-981-97-6703-8_1

insights into various biological processes at a level of detail that was previously unattainable [37, 45]. The SCS technologies are based on the ability to isolate and capture individual cells from a tissue or cell culture. Once isolated, the cells are lysed to release their DNA or RNA. This material is then amplified and sequenced using next-generation sequencing (NGS) technologies [3].

The SCS emerged as a response to the limitations of traditional bulk sequencing methods, where genetic information from a large population of cells is averaged [12]. Bulk sequencing was useful for characterizing the overall gene expression or DNA variation in a sample, but it couldn't reveal the diversity and functional differences among individual cells within that population. There are two main types of SCS: single-cell genome sequencing (scDNA-seq) and single-cell RNA sequencing (scRNA-seq). scDNA-seq sequences the entire genome of a single cell, while scRNA-seq sequences the transcriptome, which is the collection of all RNA molecules expressed by a cell at a given time [19].

The history of SCS can be summarized as follows:

- **Early 2000s**: The groundwork for SCS was laid with the development of microfluidic devices and single-cell isolation techniques. These advancements allowed researchers to physically isolate individual cells from a heterogeneous population. Development of methods to perform PCR at single-cell level which allows amplification of single-cell genomics.
- **2009–2010**: The first conceptual and technical breakthrough of single-cell RNA sequencing (scRNA-seq) method was introduced by Tang and others in 2009 [51]. This pioneering technique used microfluidics to isolate single cells and perform RNA sequencing on them. This marked the beginning of single-cell genomics.
- **2011–2012**: The Drop-seq method, which utilized droplet-based microfluidics for high-throughput scRNA-seq, was introduced. This made it possible to analyze thousands of individual cells simultaneously, enabling large-scale single-cell studies [58, 63].
- **2013-Present**: Numerous SCS methods were developed, including scDNA-seq [9], scATAC-seq [10, 46], scMethyl-seq [1], Smart-seq2 [39], and 10× Genomics (https://www.10xgenomics.com/), each with their own advantages and applications. These methods expanded the scope of single-cell analysis beyond transcriptomics, facilitating to study genomics epigenomics, etc.

1.1.2   How Does Single-Cell Sequencing Work?

Single-cell sequencing involves several key steps (Fig. 1.1).

- ***Cell Isolation***: Individual cells are physically separated from a heterogeneous population using various techniques, including fluorescence activated cell sorting (FACS), microfluidics, laser capture microdissection, microdroplets, or manual picking under a microscope.

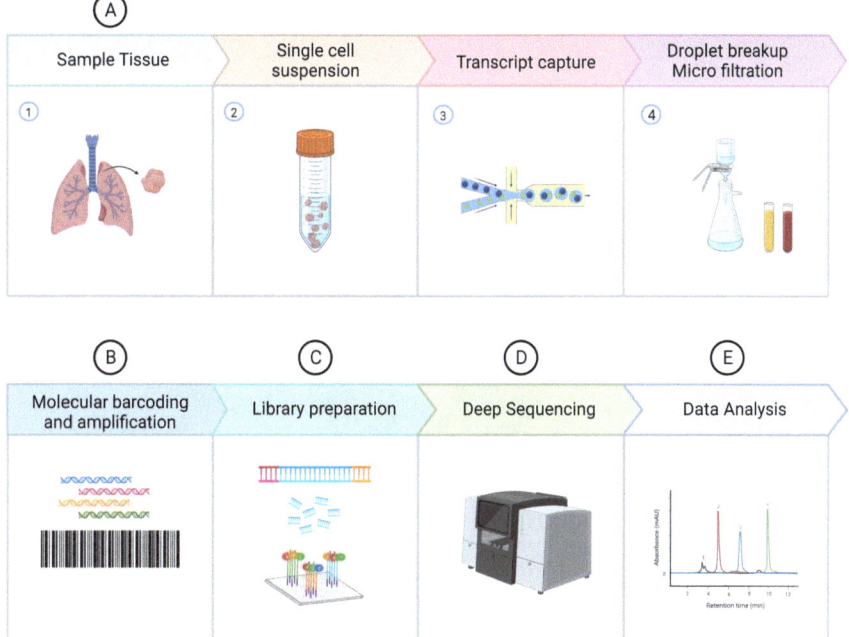

Fig. 1.1 Key steps of single-cell sequencing

- *Library Preparation*: After isolation, the cell's genetic material (DNA or RNA) is extracted, and libraries are prepared. This involves amplifying the genetic material and adding unique molecular barcodes to each cell's genetic material to distinguish them during sequencing.
- *Sequencing*: The prepared libraries are then subjected to high-throughput deep sequencing using technologies like NGS.
- *Data Analysis*: The resulting sequencing data is processed and analyzed using bioinformatics tools to determine the genetic information of each individual cell. Researchers can study gene expression, identify genetic mutations, or investigate other genomic features on a cell-by-cell basis.

1.1.3 Applications

Single-cell sequencing has numerous applications in various fields of biology and medicine, as shown in Fig. 1.2.

- **Cancer Research**: It can identify rare tumor cell subpopulations, uncover genetic heterogeneity within tumors, and understand the evolution of cancer cells.

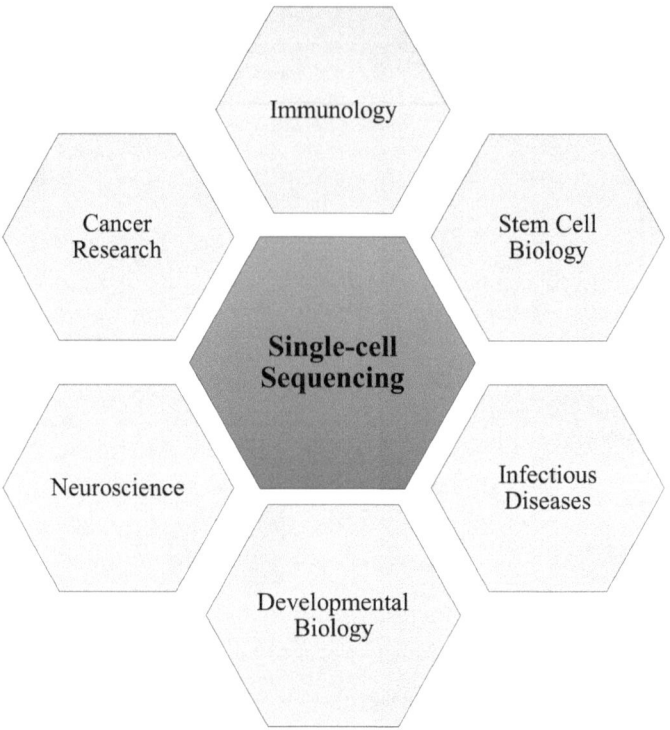

Fig. 1.2 Diverse applications of single-cell sequencing technologies

- **Neuroscience**: Single-cell sequencing helps characterize neuronal diversity and study brain development and diseases.
- **Immunology**: It aids in profiling immune cell populations, studying immune responses, and understanding autoimmune diseases.
- **Developmental Biology**: Researchers can track the differentiation of embryonic cells and study organ development at the single-cell level.
- **Stem Cell Biology**: It allows the characterization of pluripotent stem cells and their differentiation pathways.
- **Infectious Diseases**: Single-cell sequencing can be used to study host–pathogen interactions and immune responses during infections.

1.2   Overview and Significance of Single-Cell RNA-seq

Single-cell RNA sequencing (scRNA-seq) is a technique that allows scientists to measure the expression of all genes in a single cell, and lets us analyze the gene expression patterns of individual cells within a heterogeneous population. Unlike traditional bulk RNA sequencing, which provides an average gene expression profile for a whole tissue or cell population, scRNA-seq provides detailed information about the gene expression of each individual cell [24]. This technology has revolutionized our understanding of cell heterogeneity, cell types, and cellular responses to various conditions. For instance, scRNA-seq has been used to identify new cell types in the brain, to track the development of immune cells, and to understand how cancer cells evolve.

The scRNA-seq analysis holds significant importance in the field of molecular biology and has broad-ranging applications. Its significance lies in its ability to provide detailed insights into gene expression at the single-cell level, enabling researchers to study cellular heterogeneity, identify novel cell types, and understand various biological processes [14]. Here are some key significances of scRNA-seq analysis (Fig. 1.3).

- **Cell Type Identification**: scRNA-seq allows researchers to categorize and define distinct cell types within a complex tissue or organ based on their gene expression profiles. This has been particularly valuable in characterizing cell types in the brain, immune system, and other tissues.
- **Cellular Heterogeneity**: It helps uncover cellular heterogeneity within seemingly homogeneous populations, such as cancer cells. Researchers can identify rare cell subpopulations with unique gene expression signatures.
- **Development of New Diagnostic Markers**: The scRNA-seq data can be used to identify and characterize genes, gene expression patterns, and specific cellular states that can be leveraged for the diagnosis of diseases. We can identify genes that are overexpressed or underexpressed in diseased cells compared to healthy ones. These differentially expressed genes (DEGs) can serve as potential diagnostic biomarkers. For instance, in cancer research, scRNA-seq can identify unique gene expression signatures associated with different cancer subtypes or stages.
- **Target Discovery for Therapeutic Strategies**: By analyzing scRNA-seq data, we can identify genes or cellular pathways that are dysregulated in the disease. These dysregulated genes or pathways can be potential targets for therapeutic intervention. Developing drugs or therapies that specifically target these factors can lead to more effective treatments. Further, scRNA-seq can also help predict how individual cells or cell populations will respond to different treatments. By understanding the gene expression profiles of cells, researchers can assess which treatments are likely to be most effective for a specific patient or subgroup.
- **Disease Research**: In cancer research, scRNA-seq can identify and study tumor heterogeneity, drug resistance mechanisms, and rare cancer stem cell populations. It is also applied in understanding the molecular basis of various diseases, including neurodegenerative disorders and autoimmune diseases.
- **Immunology**: scRNA-seq helps characterize immune cell populations, identify cell states during immune responses, and understand immune cell function. It can be used to study autoimmune diseases, immunotherapy responses, and vaccine development.

- **Stem Cell Biology**: Researchers use scRNA-seq to investigate the properties and differentiation trajectories of stem cells, including pluripotent stem cells and adult stem cells.
- **Drug Discovery**: scRNA-seq can aid in drug discovery by identifying target genes and pathways, understanding drug responses at the single-cell level, and assessing drug toxicity.
- **Spatial Transcriptomics**: Techniques like spatial transcriptomics combine scRNA-seq data with spatial information, allowing researchers to map gene expression patterns within tissues, which is valuable for understanding spatial organization and cell–cell interactions.
- **Rare Cell Detection**: scRNA-seq can identify and study rare cell populations, such as circulating tumor cells, which are present in very low numbers and are challenging to detect using traditional methods.
- **Personalized Medicine**: With the detailed information provided by scRNA-seq, it becomes possible to design personalized treatment strategies. Physicians can use the molecular and cellular profiles of a patient's cells to choose the most appropriate therapies, potentially improving treatment outcomes and reducing side effects.
- **Understanding Cells Response**: The scRNA-seq can be used to study how cells respond to environmental conditions, stimuli, infections, drugs, and other factors. scRNA-seq allows scientists to measure the expression of all genes in a single cell, which provides a detailed snapshot of the cell's state. For instance, scRNA-seq has been used to study how cells respond to changes in temperature, pH, nutrient levels, and different stimuli, such as hormones, growth factors, and cytokines. The scRNA-seq has been used to identify genes that are upregulated or downregulated when immune cells are activated. The scRNA-seq has been used to study the response of cancer cells to chemotherapy drugs, and identified genes that are upregulated in cancer cells that are resistant to chemotherapy drugs. This information could be used to develop new drugs that target these genes and overcome chemotherapy resistance.

1.3   Challenges in scRNA-seq Data Analysis

The scRNA-seq data analysis is a complex and evolving field, and it comes with several challenges that researchers must address to obtain meaningful insights from the data. The challenges arise from the unique characteristics of scRNA-seq data, such as its high dimensionality, sparsity, and noise. Here are some of the key challenges in scRNA-seq data analysis:

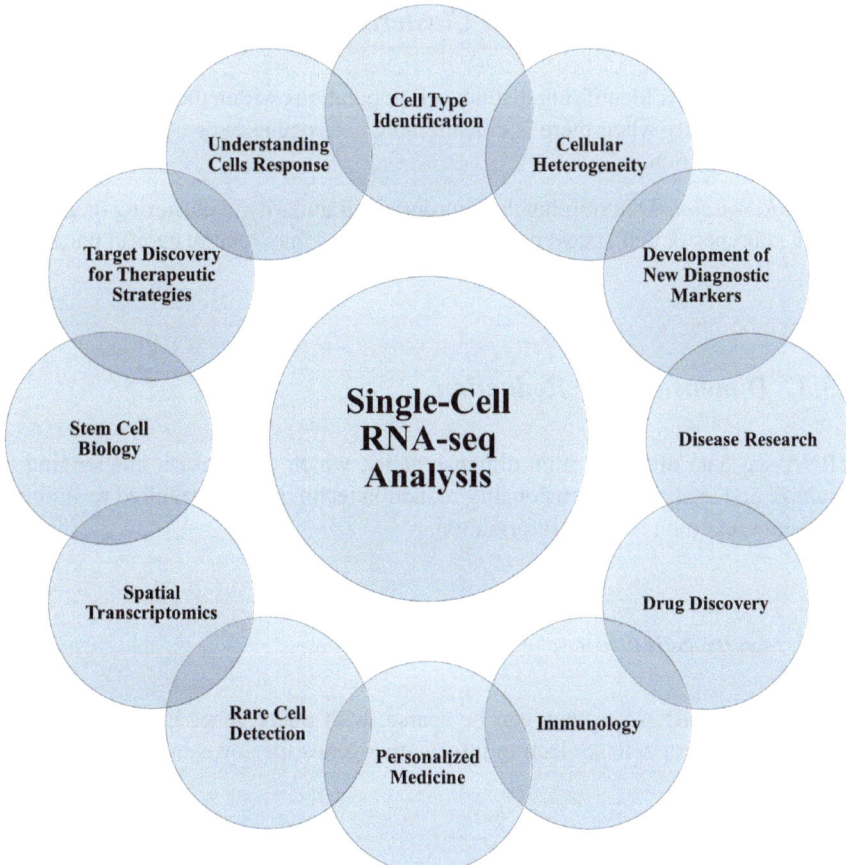

Fig. 1.3 Significance of single-cell RNA-seq analysis

1.3.1 Data Quality and Preprocessing

Data Noise: scRNA-seq data can contain technical noise, such as dropouts (genes that are not detected in some cells) and amplification bias. Proper noise handling is essential.

Batch Effects: When data is generated in different batches or experiments, batch effects can confound the analysis. Batch correction methods are necessary to remove or minimize these effects.

Normalization: Choosing the right normalization method is crucial to ensure that gene expression values are comparable across cells while preserving biological variation.

1.3.2 Cell Identification and Clustering

Cell Heterogeneity: Identifying distinct cell populations within the data can be challenging, especially when there is considerable heterogeneity or when cells exhibit continuous variation.

Cluster Resolution: Determining the appropriate granularity of clustering (e.g., identifying subtypes or cell states) can be subjective and may require careful parameter tuning.

1.3.3 Dimensionality Reduction

scRNA-seq data often has high dimensionality, which can make it challenging to visualize and analyze. Dimensionality reduction techniques are required to simplify data while retaining relevant information.

1.3.4 Gene Selection

Sparse Data: scRNA-seq data can be sparse, with many genes having low or no expression in most cells. Selecting informative genes for downstream analysis can be non-trivial.

1.3.5 Trajectory Analysis

When studying developmental processes or cell differentiation, inferring pseudotime (a measure of progression) and constructing cellular trajectories can be challenging due to multiple branching points and complex dynamics.

1.3.6 Computational Resources

Data Size: Large-scale scRNA-seq experiments generate massive datasets that require substantial computational resources for storage and analysis.

Algorithmic Efficiency: Developing and running analysis pipelines can be time-consuming, and some methods may not scale well to very large datasets.

1.3.7 Integration of Multiple Modalities

Multi-omics Integration: Integrating different types of omics data (e.g., scRNA-seq and scATAC-seq) to gain a comprehensive understanding of cell biology is challenging and requires specialized methods.

1.3.8 Biological Interpretation

Annotation: Assigning biological meaning to clusters or cell states often relies on external knowledge, and obtaining accurate annotations can be difficult.

Functional Analysis: Determining the biological functions and pathways associated with differentially expressed genes or cell states requires careful interpretation and validation.

1.3.9 Reproducibility and Benchmarking

Ensuring the reproducibility of scRNA-seq data analysis workflows and benchmarking against established datasets can be challenging due to variations in data and analysis methods.

1.3.10 User-Friendly Tools

Some of the analysis tools and software packages used in scRNA-seq analysis require a deep understanding of computational biology, making them less accessible to experimental biologists who may lack extensive computational expertise.

1.3.11 Ethical and Privacy Concerns

When working with human or patient-derived data, protecting privacy and complying with ethical standards becomes an important consideration.

1.4 Machine Learning in scRNA-seq

Machine learning (ML) is a powerful tool that plays a crucial role in the analysis of scRNA-seq) data by enabling researchers to extract meaningful insights from complex and high-dimensional datasets. ML algorithms can be used to learn from scRNA-seq data and to make predictions about the data. For example, ML algorithms are being used to develop new methods for handling the high dimensionality and sparsity of scRNA-seq data, and for identifying new cell types and studying the relationships between different cell types.

Here are various roles of machine learning in scRNA-seq data analysis (Fig. 1.4).

1.4.1 Clustering and Cell Type Identification

Clustering: ML algorithms like k-means can group cells with similar gene expression profiles into clusters, helping to identify different cell types or states within heterogeneous populations. Hierarchical clustering methods can reveal the hierarchical relationships between cell clusters, allowing for a deeper understanding of cell lineage or differentiation.

Dimensionality Reduction: Techniques like t-distributed stochastic neighbor embedding (t-SNE) and Uniform Manifold Approximation and Projection (UMAP) are used to reduce high-dimensional data into 2D or 3D representations, aiding in visualizing and interpreting cell clusters.

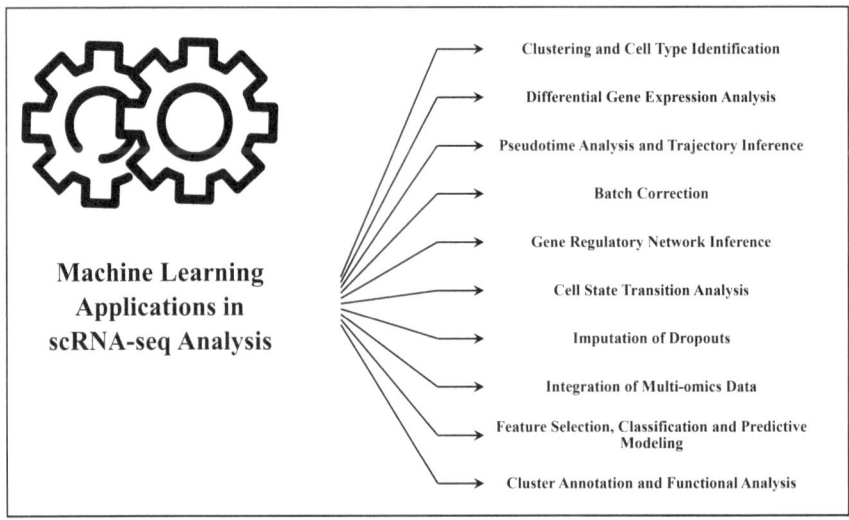

Fig. 1.4 Roles of machine learning in scRNA-seq data analysis

1.4.2   Differential Gene Expression Analysis

ML algorithms are applied to identify genes that are differentially expressed between cell clusters or conditions, helping to uncover key markers associated with specific cell types or states. Some of the tools are DESeq2, edgeR, or limma [8].

1.4.3   Pseudotime Analysis and Trajectory Inference

ML methods are employed to infer pseudotime, which represents the progression of cells along a developmental trajectory. This helps uncover the temporal order of cell states and understand cellular differentiation processes. Some of the common tools are Monocle [52], TSCAN [21], PAGA [57], and VeTra [56]. Monocle is the first pseudotime inference method that utilizes minimum spanning tree (MST) algorithm on individual cells to determine the longest path and assign the pseudotime of each cell. Another tool, The Tools for Single Cell Analysis (TSCAN) runs the MST algorithm on clusters in order to construct a cluster-based MST. Partition-based graph abstraction (PAGA) compresses and denoises original data, constructs a kNN-like graph, and finds community of vertices.

1.4.4   Batch Correction

Batch effects are the outcome of technical variation across samples which need to be prevented or corrected using computational approaches. ML algorithms correct batch effects in scRNA-seq data, ensuring that data from different experiments or batches can be integrated and analyzed without bias. Some of the common tools are Harmony [28], LIGER [54], Seurat 3 [50], ComBat [23], and MNN [15].

1.4.5   Gene Regulatory Network Inference

A gene regulatory network (GRN) represents interactions among genes and gene products, and depicted as a graph where genes are represented as nodes, and edges represent regulatory interactions. The GRNs are used to (i) understand the flow of information in a biological system, (ii) determine biological circuits used for a specific purpose, and (iii) to model gene expression changes under various conditions [43]. ML approaches can be used to infer GRNs from scRNA-seq data, helping to elucidate the regulatory interactions among genes in specific cell types or conditions. Some of the common tools are GENIE3 [20], PIDC [7], LEAP [48], and GRNBoost2 [36].

1.4.6 Cell State Transition Analysis

Cell state transitions are fundamental processes where cells change states over time, playing a crucial role in embryonic development and tissue maintenance. These transitions ensure the replacement of damaged cells during homeostasis and tissue repair. However, aberrations in these transitions can lead to various pathologies, including developmental disorders and cancers. ML methods aid in modeling and analyzing transitions between cell states within a trajectory, shedding light on the dynamics of cellular processes. Some of the common tools for cell state transition analysis are [64], Slingshot [49], CellRank [29], and sciCSR [38].

1.4.7 Imputation of Dropouts

scRNA-seq techniques are typically incapable of quantifying the expression levels of all genes in a cell, posing a demand for computational methods to predict the missing values, known as dropout imputation. ML models may impute missing gene expression values caused by dropout events in scRNA-seq data, allowing for more accurate downstream analyses. Some of the common tools for dropout imputation are scImpute [31], DrImpute [13], scHinter [59], Epi-Impute [41], and scRecover [34].

1.4.8 Integration of Multi-omics Data

ML techniques facilitate the integration of scRNA-seq data with other omics data types (e.g., scATAC-seq, scDNA-seq) to gain a comprehensive understanding of cellular biology. Some of the tools are Multi-Omics Factor Analysis (MOFA+) [4], scAI [22], and Seurat 4.0 [16].

1.4.9 Feature Selection, Classification and Predictive Modeling

One of the essential tasks in scRNA-seq data analysis is feature selection. It is generally used for gene dimension reduction, gene marker identification, and cell type classification. Traditional and the most widely used methods for feature selection in scRNA-seq data are based on a differential distribution where a statistical model (e.g. fold-change, t-test, etc.) is used to find changes in gene expression among cell types, usually called differentially expressed genes (DEGs). Recent advancements in machine learning and deep learning techniques provide ML-based feature selection methods, providing an alternative to traditional differential distribution-based methods where the importance of genes is computed by neural networks [18]. Some of the popular ML-based feature selection methods are Occlusion [61], LIME [44], DeepLIFT [47], GradientShap [33], ReliefF [53], FeatureAblation [27], and scDeepFeatures [17].

Once important gene feature selection is done, a classifier may be trained to classify and detect cell types. ML algorithms like SVM and random forests can be trained to classify cells into predefined categories, such as disease vs. healthy or treatment-responsive vs. non-responsive. Several cell-type classification and predictive modeling methods have been developed for scRNA-seq data. Some of the ML-based cell type classification methods are scmapcluster and scmapcell methods from scmap [26] package, CasTLe [32], scPred [2], and scID [5].

1.4.10 Cluster Annotation and Functional Analysis

The important objective of annotation and functional analysis is to gain biological insight. Annotating a query scRNA-seq dataset means assigning a cell type identity to each one of the query single cells, or a group of cells at once (called cluster annotation). Some of the common annotation platforms are CellMarker [62], PanglaoDB [11], CancerSEA [60]. CellMarker is a catalog of manually curated human and mouse cell-type markers that are categorized according to their tissue of origin, hierarchically grouped by localization, morphology, and functionality. PanglaoDB is a cell type atlas that stores gene expression information and its relation to cell types. CancerSEA catalogs markers of functional cell states in cancer. These databases and online repositories offer handy tools for cell-type to marker-gene relations annotation.

Several ML-based methods have been proposed for the purpose which include (semi-) supervised and unsupervised. The (semi-) supervised methods primarily utilized previously annotated datasets to annotate new datasets [30], which includes Garnett [40], MARS [6], and scPipeline [35]—a toolbox that offers modular workflows for multi-level cellular annotation.

1.5   Conclusion

scRNA-seq is a powerful technology that allows researchers to investigate gene expression profiles at the single-cell level. This approach provides a more detailed and nuanced understanding of cellular heterogeneity compared to traditional bulk RNA-seq, where the average expression of genes in a population of cells is measured. Machine learning plays a significant role in analyzing scRNA-seq data by enabling the extraction of meaningful patterns, identification of cell types, and interpretation of complex biological information. This chapter briefly presents an overview, significance, and challenges of scRNA-seq data analysis. Further, the role of machine learning in scRNA-seq data analysis has been discussed in length, including clustering and cell-type identification, differential gene expression analysis, pseudotime analysis and trajectory inference, batch correction, gene regulation network inference, cell state transition analysis, imputation of dropouts, integration of multi-omics data, feature selection, classification, and predicting modeling, and annotation and function analysis.

References

1. J. Ahn, S. Heo, J. Lee, D. Bang, Introduction to single-cell DNA methylation profiling methods. Biomolecules **11**(7), 1013 (2021)
2. J. Alquicira-Hernandez, A. Sathe, H.P. Ji, Q. Nguyen, J.E. Powell, scPred: accurate supervised method for cell-type classification from single-cell RNA-seq data. Genome Biol. **20**(1), 1–17 (2019)
3. N. Anaparthy, Y.J. Ho, L. Martelotto, M. Hammell, J. Hicks, Single-cell applications of next-generation sequencing. Cold Spring Harbor Perspect. Med. **9**(10) (2019)
4. R. Argelaguet, D. Arnol, D. Bredikhin, Y. Deloro, B. Velten, J.C. Marioni, O. Stegle, MOFA+: a statistical framework for comprehensive integration of multi-modal single-cell data. Genome Biol. **21**(1), 1–17 (2020)
5. K. Boufea, S. Seth, N.N. Batada, scID uses discriminant analysis to identify transcriptionally equivalent cell types across single-cell RNA-seq data with batch effect. iScience **23**(3) (2020). Code available on GitHub: https://github.com/BatadaLab/scID
6. M. Brbić, M. Zitnik, S. Wang, A.O. Pisco, R.B. Altman, S. Darmanis, J. Leskovec, MARS: discovering novel cell types across heterogeneous single-cell experiments. Nat. Methods **17**(12), 1200–1206 (2020)
7. T.E. Chan, M.P.H. Stumpf, A.C. Babtie, Gene regulatory network inference from single-cell data using multivariate information measures. Cell Syst. **5**(3), 251–267. e3 (2017)
8. S. Das, A. Rai, S.N. Rai, Differential expression analysis of single-cell RNA-seq data: current statistical approaches and outstanding challenges. Entropy **24**(7), 995 (2022)
9. G.D. Evrony, A.G. Hinch, C. Luo, Applications of single-cell DNA sequencing. Annu. Rev. Genomics Hum. Genet. **22**, 171–197 (2021)
10. R. Fang, S. Preissl, Y. Li, X. Hou, J. Lucero, X. Wang, B. Ren et al., Comprehensive analysis of single cell ATAC-seq data with SnapATAC. Nat. Commun. **12**(1), 1337 (2021)
11. O. Franzén, L.M. Gan, J.L. Björkegren, PanglaoDB: a web server for exploration of mouse and human single-cell RNA sequencing data. Database, baz046 (2019)
12. C. Gawad, W. Koh, S.R. Quake, Single-cell genome sequencing: current state of the science. Nat. Rev. Genet. **17**(3), 175–188 (2016)
13. W. Gong, I.Y. Kwak, P. Pota, N. Koyano-Nakagawa, D.J. Garry, DrImpute: imputing dropout events in single cell RNA sequencing data. BMC Bioinform. **19**, 1–10 (2018)
14. I.N. Grabski, K. Street, R.A. Irizarry, Significance analysis for clustering with single-cell RNA-sequencing data. Nat. Methods **20**(8), 1196–1202 (2023)
15. L. Haghverdi, A.T. Lun, M.D. Morgan, J.C. Marioni, Batch effects in single-cell RNA-sequencing data are corrected by matching mutual nearest neighbors. Nat. Biotechnol. **36**(5), 421–427 (2018)
16. Y. Hao, S. Hao, E. Andersen-Nissen, W.M. Mauck, S. Zheng, A. Butler, R. Satija et al., Integrated analysis of multimodal single-cell data. Cell **184**(13), 3573–3587 (2021)
17. H. Huang, P. Yang, ScDeepFeatures: deep learning-based feature selection for single-cell RNA sequencing data analysis. Zenodo (2023). https://doi.org/10.5281/zenodo.10027186
18. H. Huang, C. Liu, M.M. Wagle, P. Yang, Evaluation of deep learning-based feature selection for single-cell RNA sequencing data analysis. Genome Biol. **24**(1), 259 (2023)

19. X. Huang, S. Liu, L. Wu, M. Jiang, Y. Hou, High throughput single cell RNA sequencing, bioinformatics analysis and applications, in *Single Cell Biomedicine* (2018), pp. 33–43
20. V.A. Huynh-Thu, A. Irrthum, L. Wehenkel, P. Geurts, Inferring regulatory networks from expression data using tree-based methods. PLoS ONE **5**(9), e12776 (2010)
21. Z. Ji, H. Ji, TSCAN: pseudo-time reconstruction and evaluation in single-cell RNA-seq analysis. Nucleic Acids Res. **44**(13), e117–e117 (2016)
22. S. Jin, L. Zhang, Q. Nie, ScAI: an unsupervised approach for the integrative analysis of parallel single-cell transcriptomic and epigenomic profiles. Genome Biol. **21**, 1–19 (2020)
23. W.E. Johnson, C. Li, A. Rabinovic, Adjusting batch effects in microarray expression data using empirical Bayes methods. Biostatistics **8**(1), 118–127 (2007)
24. D. Jovic, X. Liang, H. Zeng, L. Lin, F. Xu, Y. Luo, Single-cell RNA sequencing technologies and applications: a brief overview. Clin. Transl. Med. **12**(3), e694 (2022)
25. Y. Kashima, Y. Sakamoto, K. Kaneko, M. Seki, Y. Suzuki, A. Suzuki, Single-cell sequencing techniques from individual to multiomics analyses. Exp. Mol. Med. **52**(9), 1419–1427 (2020)
26. V.Y. Kiselev, A. Yiu, M. Hemberg, Scmap: projection of single-cell RNA-seq data across data sets. Nat. Methods **15**(5), 359–362 (2018)
27. N. Kokhlikyan, V. Miglani, M. Martin, E. Wang, B. Alsallakh, J. Reynolds, O. Reblitz-Richardson, et al.: Captum: a unified and generic model interpretability library for pytorch (2020). arXiv preprint arXiv:2009.07896
28. I. Korsunsky, N. Millard, J. Fan, K. Slowikowski, F. Zhang, K. Wei, S. Raychaudhuri et al., Fast, sensitive and accurate integration of single-cell data with Harmony. Nat. Methods **16**(12), 1289–1296 (2019)
29. M. Lange, V. Bergen, M. Klein, M. Setty, B. Reuter, M. Bakhti, F.J. Theis et al., Cell rank for directed single-cell fate mapping. Nat. Methods **19**(2), 159–170 (2022)
30. D. Li, J. Ding, Z. Bar-Joseph, Unsupervised cell functional annotation for single-cell RNA-seq. Genome Res. **32**(9), 1765–1775 (2022)
31. W.V. Li, J.J. Li, An accurate and robust imputation method scImpute for single-cell RNA-seq data. Nat. Commun. **9**(1), 997 (2018)
32. Y. Lieberman, L. Rokach, T. Shay, CaSTLe–classification of single cells by transfer learning: harnessing the power of publicly available single cell RNA sequencing experiments to annotate new experiments. PLoS ONE **13**(10), e0205499 (2018)
33. S.M. Lundberg, S.I. Lee, A unified approach to interpreting model predictions, in *Advances in Neural Information Processing Systems*, vol. 30 (2017)
34. Z. Miao, X. Zhang, scRecover: scRecover for imputation of single-cell RNA-seq data. R package version 1.18.0 (2023). https://bioconductor.org/packages/scRecover
35. N. Mikolajewicz, R. Gacesa, M. Aguilera-Uribe, K.R. Brown, J. Moffat, H. Han, Multi-level cellular and functional annotation of single-cell transcriptomes using scPipeline. Commun. Biol. **5**(1), 1142 (2022)
36. T. Moerman, S. Aibar Santos, C. Bravo González-Blas, J. Simm, Y. Moreau, J. Aerts, S. Aerts, GRNBoost2 and arboreto: efficient and scalable inference of gene regulatory networks. Bioinformatics **35**(12), 2159–2161 (2019)
37. T. Nawy, Single-cell sequencing. Nat. Methods **11**(1), 18–18 (2014)
38. J.C. Ng, G. Montamat Garcia, A.T. Stewart, P. Blair, C. Mauri, D.K. Dunn-Walters, F. Fraternali, sciCSR infers B cell state transition and predicts class-switch recombination dynamics using single-cell transcriptomic data. Nat. Methods, 1–12 (2023)
39. S. Picelli, O.R. Faridani, Å.K. Björklund, G. Winberg, S. Sagasser, R. Sandberg, Full-length RNA-seq from single cells using Smart-seq2. Nat. Protoc. **9**(1), 171–181 (2014)
40. H.A. Pliner, J. Shendure, C. Trapnell, Supervised classification enables rapid annotation of cell atlases. Nat. Methods **16**(10), 983–986 (2019)
41. M. Raevskiy, V. Yanvarev, S. Jung, A. Del Sol, Y.A. Medvedeva, Epi-impute: single-cell RNA-seq imputation via integration with single-cell ATAC-seq. Int. J. Mol. Sci. **24**(7), 6229 (2023)
42. K. Raza, S. Ahmad, Recent advancement in next-generation sequencing techniques and its computational analysis. Int. J. Bioinform. Res. Appl. **15**(3), 191–220 (2019)

43. K. Raza, Fuzzy logic based approaches for gene regulatory network inference. Artif. Intell. Med. **97**, 189–203 (2019)
44. M.T. Ribeiro, S. Singh, C. Guestrin, "Why should i trust you?" Explaining the predictions of any classifier, in *Proceedings of the 22nd ACM SIGKDD International Conference on Knowledge Discovery and Data Mining* (2016), pp. 1135–1144
45. E. Shapiro, T. Biezuner, S. Linnarsson, Single-cell sequencing-based technologies will revolutionize whole-organism science. Nat. Rev. Genet. **14**(9), 618–630 (2013)
46. P. Shi, Y. Nie, J. Yang, W. Zhang, Z. Tang, J. Xu, Fundamental and practical approaches for single-cell ATAC-seq analysis. Abiotech **3**(3), 212–223 (2022)
47. A. Shrikumar, P. Greenside, A. Kundaje, Learning important features through propagating activation differences, in *International Conference on Machine Learning* (PMLR, 2017), pp. 3145–3153
48. A.T. Specht, J. Li, LEAP: constructing gene co-expression networks for single-cell RNA-sequencing data using pseudotime ordering. Bioinformatics **33**(5), 764–766 (2017)
49. K. Street, D. Risso, R.B. Fletcher, D. Das, J. Ngai, N. Yosef, S. Dudoit et al., Slingshot: cell lineage and pseudotime inference for single-cell transcriptomics. BMC Genom. **19**, 1–16 (2018)
50. T. Stuart, A. Butler, P. Hoffman, C. Hafemeister, E. Papalexi, W.M. Mauck, R. Satija et al., Comprehensive integration of single-cell data. Cell **177**(7), 1888–1902 (2019)
51. F. Tang, C. Barbacioru, Y. Wang, E. Nordman, C. Lee, N. Xu, M.A. Surani et al., MRNA-seq whole-transcriptome analysis of a single cell. Nat. Methods **6**(5), 377–382 (2009)
52. C. Trapnell, D. Cacchiarelli, J. Grimsby, P. Pokharel, S. Li, M. Morse, J.L. Rinn et al., The dynamics and regulators of cell fate decisions are revealed by pseudotemporal ordering of single cells. Nat. Biotechnol. **32**(4), 381–386 (2014)
53. R.J. Urbanowicz, M. Meeker, W. La Cava, R.S. Olson, J.H. Moore, Relief-based feature selection: Introduction and review. J. Biomed. Inform. **85**, 189–203 (2018)
54. J. Welch, V. Kozareva, A. Ferreira, C. Vanderburg, C. Martin, E. Macosko, Integrative inference of brain cell similarities and differences from single-cell genomics. BioRxiv, 459891 (2018)
55. L. Wen, F. Tang, Recent advances in single-cell sequencing technologies. Precis. Clin. Med. **5**(1), pbac002 (2022)
56. G. Weng, J. Kim, K.J. Won, VeTra: a tool for trajectory inference based on RNA velocity. Bioinformatics **37**(20), 3509–3513 (2021)
57. F.A. Wolf, F.K. Hamey, M. Plass, J. Solana, J.S. Dahlin, B. Göttgens, F.J. Theis et al., PAGA: graph abstraction reconciles clustering with trajectory inference through a topology preserving map of single cells. Genome Biol. **20**, 1–9 (2019)
58. Z. Xu, Y. Wang, K. Sheng, R. Rosenthal, N. Liu, X. Hua, Y. Wang et al., Droplet-based high-throughput single microbe RNA sequencing by smRandom-seq. Nat. Commun. **14**(1), 5130 (2023)
59. P. Ye, W. Ye, C. Ye, S. Li, L. Ye, G. Ji, X. Wu, scHinter: imputing dropout events for single-cell RNA-seq data with limited sample size. Bioinformatics **36**(3), 789–797 (2020)
60. H. Yuan, M. Yan, G. Zhang, W. Liu, C. Deng, G. Liao, X. Li et al., CancerSEA: a cancer single-cell state atlas. Nucleic Acids Res. **47**(D1), D900–D908 (2019)
61. M.D. Zeiler, R. Fergus, Visualizing and understanding convolutional networks, in *Computer Vision–ECCV 2014: 13th European Conference, Zurich, Switzerland, September 6–12, 2014, Proceedings, Part I 13* (Springer International Publishing, 2014), pp. 818–833
62. X. Zhang, Y. Lan, J. Xu, F. Quan, E. Zhao, C. Deng, Y. Xiao et al., Cell marker: a manually curated resource of cell markers in human and mouse. Nucleic Acids Res. **47**(D1), D721–D728 (2019)
63. W.M. Zhou, Y.Y. Yan, Q.R. Guo, H. Ji, H. Wang, T.T. Xu, J.Y. Zhang et al., Microfluidics applications for high-throughput single cell sequencing. J. Nanobiotechnol. **19**, 1–21 (2021)
64. G. Zhu, H. Yang, X. Chen, J. Wu, Y. Zhang, X.M. Zhao, CSTEA: a webserver for the cell state transition expression atlas. Nucleic Acids Res. **45**(W1), W103–W108 (2017)

Chapter 2
Preprocessing and Quality Control

2.1 Background

Recent advancements in scRNA-seq technologies led to high-quality runs with high throughputs, however, scRNA-seq datasets contain systematic and random noise. Therefore, data preprocessing and quality control (QC) become an essential step in the analysis of scRNA-seq data [12]. It involves a series of steps and techniques to clean, normalize, and transform the raw data generated from scRNA-seq experiments into a format suitable for downstream analysis. Data preprocessing aims to reduce noise, correct technical artifacts, and make the data more amenable to statistical analysis and interpretation [18]. Further, QC of scRNA-seq data is also important before performing downstream analysis and making any biological conclusions. The major goal of QC analysis is to assess metrics related to the quality of data and remove poor-quality data and noise.

2.2 Quality Control and Metrics

Quality control (QC) is one of the important steps in scRNA-seq data preprocessing whose primary purpose is to assess the quality of the data generated from scRNA-seq experiments, identify and filter out poor-quality cells, and ensure that the data is reliable for downstream analysis [13].

QC metrics are a set of quantitative measures and criteria used to assess the quality and reliability of the scRNA-seq data generated from individual cells. These metrics help researchers identify potential issues, such as poor-quality cells, filtering doublets, and batch effects, and make informed decisions about data preprocessing and filtering. Some of the common QC steps and metrics used in scRNA-seq are described as follows.

K. Raza, *Machine Learning in Single-Cell RNA-seq Data Analysis*,
SpringerBriefs in Computational Intelligence,
https://doi.org/10.1007/978-981-97-6703-8_2

2.2.1 Assessment of Cell Quality

Low-quality cells mean those cells having low sequencing depth, low gene counts, or high mitochondrial gene content (indicating cell stress or damage) that need to be excluded from the analysis.

Sequencing Depth: Check the sequencing depth for each cell, which is the number of reads or UMIs (unique molecular identifiers) mapped to that cell. Cells with very low sequencing depth may not provide reliable information and can be excluded. The threshold for very low sequencing depth that should be used for exclusion in scRNA-seq data analysis varies depending on several factors, including the experimental design, the specific goals of the analysis, and the characteristics of the dataset. There is no universally agreed-upon threshold, and the choice often depends on the context of your study [23].

Gene Count: Examine the number of genes detected in each cell. Cells with very few detected genes may be low-quality and should be filtered out.

Mitochondrial Gene Content: Calculate the proportion of reads mapping to mitochondrial genes in each cell. High mitochondrial gene content can be indicative of cell stress or damage and may suggest poor-quality cells [5].

Further, we can also visualize QC metrics, such as sequencing depth, gene count, and mitochondrial gene content, using plots like histograms, scatterplots, or violin plots. This can help identify patterns and outliers.

2.2.2 Identification and Removal of Doublets

Doublets are cells that contain genetic material from two or more cells. They can be identified based on their unique gene expression patterns, which often show higher expression of genes from multiple cell types. Software tools like DoubletFinder [22], Scrublet [28], Solo [3], doubletD [27], and SoCube [30] are commonly used to detect and remove doublets from scRNA-seq data.

2.2.3 Batch Effects and Technical Artifacts

Check for batch effects if the data comes from multiple experimental batches. Batch effects can be corrected using batch correction techniques such as Seurat v5 integration or Harmony [16], and JIVE [11]. Correct for technical artifacts introduced during library preparation or sequencing using methods like scran [19], scater [21], MNN [8], and CellBender [6].

2.3 Preprocessing

2.3.1 Normalization

Removal of technical biases: Data may be corrected for systematic biases introduced during library preparation and sequencing, such as batch effects, sequencing depth differences, and gene length biases.

Scaling: Gene expression values may be scaled to make them comparable across cells, often by transforming them into log-transformed counts per million (log-CPM) or transcripts per million (TPM) [1, 20].

2.3.2 Gene Filtering

Gene filtering in scRNA-seq data analysis is a preprocessing step that involves selecting a subset of genes for downstream analysis while excluding genes that may introduce noise or do not provide meaningful information. Gene filtering is essential for reducing the dimensionality of the data, improving computational efficiency, and focusing on genes that are biologically relevant to the research question [10]. For instance, lowly expressed genes and genes with low variance across cells can introduce noise into the analysis. Filtering them out can improve the signal-to-noise ratio (SNR). The common gene filtering criteria are [9]:

Expression Threshold: Genes expressed below a certain threshold (e.g., a minimum count or TPM value) across all cells may be filtered out as they may not contribute significantly to the analysis.

Variance Threshold: Genes with low variance across cells may be excluded, as they are less likely to capture meaningful biological differences.

Fraction of Expressed Cells: Genes that are expressed in only a small fraction of cells may be filtered out, as they may represent rare and potentially noisy transcripts.

Predefined Gene Sets: Some studies focus on a predefined set of genes of interest, such as marker genes for specific cell types or biological pathways.

Some of the software tools used for gene filtering in scRNA-seq are Seurat, Scanpy, Scater, edgeR, DESeq2, or Custom Scripts (own developed scripts).

2.3.3 Batch Correction

Batch effects are systematic variations in data that arise from non-biological sources, such as differences in experimental conditions, equipment settings, reagents, or personnel. These variations can confound the interpretation of experimental results and lead to incorrect conclusions if not properly addressed. Batch effects are a common concern in various high-throughput biological experiments, including scRNA-seq.

For instance, suppose you are conducting a scRNA-seq experiment to study gene expression in immune cells from different patients. You collect cells from three different individuals (Patients A, B, and C) and process them on three different days (Batch 1, Batch 2, and Batch 3) in the laboratory. Each batch is subjected to the same scRNA-seq protocol, and the resulting data is merged for analysis. The challenge arises when you analyze the merged data. You observe that there are differences in gene expression patterns between the cells from different batches. These differences are not due to biological variation but instead arise from the different processing conditions on different days. These systematic variations in gene expression are what we refer to as batch effects.

If the scRNA-seq data comes from multiple experimental batches or conditions, batch effects can be corrected to ensure that differences between batches do not confound biological signals. To mitigate batch effects in scRNA-seq and other experiments, various strategies can be employed such as batch correction methods, experimental design, quality control, and normalization [25]. In batch correction methods, several computational methods, such as ComBat-seq [31] and Harmony [16], are available to adjust for batch effects by transforming the data to remove systematic variations introduced by different batches. In experimental design, careful experimental design can minimize batch effects. In quality control, implementing QC measures, such as assessing QC metrics for each batch separately and removing problematic cells or genes, can improve data quality. Further, applying appropriate normalization techniques can reduce the impact of batch effects by scaling the data to a common reference.

2.3.4 Dimensionality Reduction

scRNA-seq data typically involves thousands of genes measured across individual cells, resulting in a high-dimensional space. Dimensionality reduction methods aim to reduce the number of dimensions (genes) while preserving the underlying biological variation [29]. This reduction is important for data visualization, clustering, and downstream analysis. There are several methods for dimensionality reduction in scRNA-seq data, few are discussed as follows:

Principal Component Analysis (PCA): PCA is one of the most commonly used dimensionality reduction techniques in scRNA-seq analysis. It identifies linear combinations of genes (principal components) that capture the most variation in the data. By selecting a subset of the top principal components, you can reduce the data to a lower-dimensional space for visualization and clustering. PCA is implemented in various scRNA-seq analysis packages like Seurat and Scanpy.

t-Distributed Stochastic Neighbor Embedding (t-SNE): t-SNE is a nonlinear dimensionality reduction technique that is particularly useful for visualizing scRNA-seq data in two or three dimensions. It aims to represent similar cells (based on gene expression patterns) close to each other in the reduced space. t-SNE is excellent for exploring and visualizing cell clusters but doesn't always preserve global relationships [15].

Uniform Manifold Approximation and Projection (UMAP): UMAP is another nonlinear dimensionality reduction technique that is gaining popularity in scRNA-seq analysis. It shares similarities with t-SNE but offers computational advantages and can capture both local and global structures in the data. UMAP is becoming a preferred choice for visualizing scRNA-seq data [2, 4].

Diffusion Maps: Diffusion maps are based on the concept of random walks on a graph constructed from the data. They capture the intrinsic geometry of the data by modeling how information diffuses through it. Diffusion maps can be used for dimensionality reduction and visualization of scRNA-seq data [7].

Non-negative Matrix Factorization (NMF): NMF is a technique that factors the gene expression matrix into two lower-dimensional matrices representing basis vectors and coefficients. NMF can uncover interpretable gene expression patterns and is used for dimensionality reduction and clustering in scRNA-seq analysis [14].

Independent Component Analysis (ICA): ICA is a technique that aims to separate the observed data into statistically independent components. It has been used in some scRNA-seq analysis pipelines to identify independent biological signals within the data [26].

2.4 Software Tools for Data Preprocessing and QC

Seurat: Seurat is a widely used R package for scRNA-seq analysis that includes functions for quality control, data normalization, dimensionality reduction, and more.

Scrublet: Scrublet is a Python package specifically designed for doublet detection in scRNA-seq data.

SingleR: SingleR is a tool for cell type annotation, which can be used as part of QC to assess the quality of cell type assignments.

Scanpy: Scanpy is an open-source Python library for single-cell analysis, including QC, preprocessing, clustering, and visualization.

Cell Ranger: Cell Ranger is a set of tools developed by $10\times$ Genomics for processing and analyzing scRNA-seq data, including QC steps.

FastQC: While not specific to scRNA-seq, FastQC is a general-purpose tool for assessing the quality of sequencing data and can be used for initial QC checks.

2.5 Case Study: Preprocessing and QC Using Python and Scanpy

This section presents a case study of scRNA-seq data preprocessing and QC using Python and Scanpy, a popular package for scRNA-seq analysis. In this case study, we'll go through the main steps involved in preprocessing and QC using hypothetical dataset of single-cell gene expression.

Background: Suppose a researcher is interested in studying gene expression in immune cells from three different patients, each processed on different days. You have collected scRNA-seq data from 10,000 cells, and your goal is to preprocess the data, perform QC, and identify cell populations.

Example dataset: The **scRNA_seq_data.h5ad** file mentioned in the code example is typically a Hierarchical Data Format 5 (HDF5) AnnData file used in scRNA-seq analysis with the Scanpy library. This file format is commonly used to store and share single-cell RNA sequencing data, including gene expression matrices, cell metadata, and various annotations. Figure 2.1 presents a simplified representation of the structure and contents of an HDF5 AnnData file for scRNA-seq data:

The snapshot of what the data in the scRNA_seq_data.h5ad file might look like when viewed in a simplified text format is depicted in Fig. 2.2.

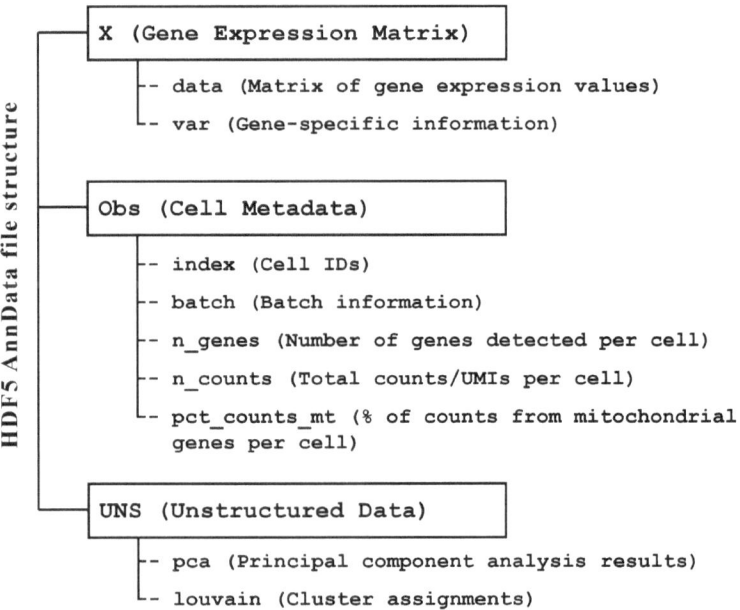

Fig. 2.1 A typical structure of an HDF5 AnnData file

```
# X (Gene Expression Matrix)
# Rows: Genes, Columns: Cells
```

Gene	Cell1	Cell2	Cell3	...
Gene1	9.6	9.0	1.0	...
Gene2	8.6	9.2	3.4	...
Gene3	3.1	5.0	7.0	...
...	...	...	...	...

```
# Obs (Cell Metadata)
# Each row represents cell, and columns contain
  cell-specific information
```

Cell_ID	Batch	n_genes	n_counts	pct_counts_mt
Cell1	Batch1	1542	12854	4.5
Cell2	Batch2	786	8700	8.5
Cell3	Batch1	905	10544	6.8
...	...	...	...	...

```
# UNS (Unstructured Data)
# Contains additional analysis results
  (e.g. PCA and clustering)
```

Fig. 2.2 Simplified example of scRNA-seq data in HDF5 AnnData format

In simplified example (Fig. 2.2), the X section of the file represents the gene expression matrix, where each row corresponds to a gene, and each column represents a cell, with values indicating gene expression levels. The Obs section contains cell metadata, including batch information and PC metrics like the number of detected genes, total counts, and the percentage of counts from mitochondrial genes. The UNS section may contain unstructured data used for further analysis, such as PCA results or cluster assignments. In a real-world scRNA-seq experiment, the scRNA-seq data file would contain actual gene expression data, and the dimensions of the matrix would be much larger. Researchers can use libraries like Scanpy in Python to read, process, and analyze data stored in this format efficiently. We need to install all the related python libraries.

```
pip install harmony
pip install --upgrade scanpy
```

Step 1: Data Import and Preprocessing

Data Import: You start by importing the scRNA-seq data into Python. The data consists of a matrix where rows represent genes and columns represent individual cells, usually in h5ad or h5 format. The data may be loaded from local file system, or from a URL.

```
# Loading scRNA-seq data from local file system
import scanpy as sc
adata = sc.read('scRNA_seq_data.h5ad')
```

```
# Loading .h5ad format scRNA-seq data from a URL
import scanpy as sc
import requests
from io import BytesIO
# Load data in h5ad format
url = 'https://figshare.com/ndownloader/files/39347357'
# Download the .h5ad file
response = requests.get(url)
# Check if request was successful
response.raise_for_status()
# Read the downloaded content into an AnnData object
adata = sc.read_h5ad(BytesIO(response.content))
```

```
# Loading 10x data in .h5 format
import scanpy as sc
# Replace data.h5 filename and URL with actual one
adata= sc.read_10x_h5(filename="data.h5",
                      backup_url="URL",)
```

Data Preprocessing: Next, you preprocess the data by performing the following steps:

* Making some of the variables (or genes) unique: Once the data is loaded, scanpy may show a warning message that not all variable names are unique, indicating that some variables or genes appear more than once. The function var_names_

make_unique() makes the variable name unique by appending a number string to each duplicate index element.

- Gene filtering: It is used to remove lowly expressed and low-variance genes.
- Log transformation to stabilize variance: This is especially useful for scRNA-seq data.
- Normalization to account for differences in sequencing depth and library size.
- Batch correction to address batch effects.

```
# Gene filtering
# Remove genes expressed in fewer than 1 cell
sc.pp.filter_genes(adata, min_counts=1)
# Log transformation and normalization
sc.pp.log1p(adata)
sc.pp.normalize_total(adata)
# Batch correction (using Harmony)
sc.pp.harmony(adata, key='batch')
```

Step 2: Quality Control (QC)

QC Metrics: The first step of QC is to compute covariates or metric, which can be calculated using the scanpy function `sc.pp.calculate_qc_metrics()`. It can also compute the proportions of counts for specific gene populations such as mitochondrial, ribosomal and hemoglobin genes. The mitochondrial counts are annotated either with the prefix "mt-" or "MT-" depending on the species of the considered data. We have used Kuppe Visium Human Heart 2022 Control datasets for the purpose of demonstration which is available on https://doi.org/10.6084/m9.fig share.22132958.v1 [24], which belongs to the work by Kuppe et al. [17]. It contains data from human cardiac remodelling after myocardial infarction using single-cell gene expression, chromatin accessibility and spatial transcriptomic profiling from patients with myocardial infarction and controls. As data belong to human, mitochondrial counts are annotated with the prefix "MT-". For mouse datasets, the prefix is usually lower case, so "mt-". Common QC metrics include:

- The percentage of mitochondrial, ribosomal, and hemoglobin genes in each cell.
- The number of genes detected per cell.
- The number of UMIs per cell.
- Batch labels to visualize batch effects.

```
# Calculate QC metrics
# mitochondrial genes
adata.var["mt"] = adata.var_names.str.startswith("MT-")
# ribosomal genes
adata.var["ribo"] = adata.var_names.str.startswith(("RPS",
"RPL"))
```

```
# hemoglobin genes
adata.var["hb"] = adata.var_names.str.contains(("^HB[^(P)]"))
sc.pp.calculate_qc_metrics(adata,   qc_vars=["mt",   "ribo",
"hb"], inplace=True, percent_top=[20], log1p=True)
p1   =   sns.displot(adata.obs["total_counts"],   bins=100,
kde=False)
p2 = sc.pl.violin(adata, "pct_counts_mt")
p3   =   sc.pl.scatter(adata,   "total_counts",   "n_genes_by_
counts", color="pct_counts_mt")
```

In the above code example, n_genes_by_counts in.obs represents the number of genes with positive counts in a cell, total_counts is the total number of counts for a cell (also known as library size), and pct_counts_mt is the proportion of total counts for a cell which are mitochondrial. Further, we plotted three QC covariates n_genes_by_counts, total_counts and pct_counts_mt per sample to assess how well the respective cells were captured (Fig. 2.3a–c). The plots indicate that some reads have higher percent of mitochondrial counts (pct_counts_mt), usually associated with cell degradation. The number of counts per cell is sufficiently high and but pct_counts_mt reads for most cells is less than 50%, we need to filter it.

Cell Filtering: Based on QC metrics, we may filter out low-quality cells to improve data quality. In the following code example, we filtered cells with more than 500 genes, and less than 5000 genes, and cells with less than 10% mitochondrial content. After applying filter, we have replotted the metrics (Fig. 2.4).

```
# Filter cells based on QC metrics
# Cells with more than 500 genes
adata = adata[adata.obs['n_genes_by_counts'] > 500]
# Cells with fewer than 5000 genes
adata = adata[adata.obs['n_genes_by_counts'] < 5000]
# Cells with less than 10% mitochondrial content
adata=adata[adata.obs['pct_counts_mt'] < 10]
# Replotting the metrics
p1   =   sc.pl.scatter(adata,   "total_counts",   "n_genes_by_
counts", color="pct_counts_mt")
```

Step 3: Dimensionality Reduction and Clustering

Dimensionality Reduction: You perform dimensionality reduction using principal component analysis (PCA) and visualize the data (Fig. 2.5).

```
# Perform PCA
sc.pp.pca(adata, n_comps=3)
# Visualize PCA results
```

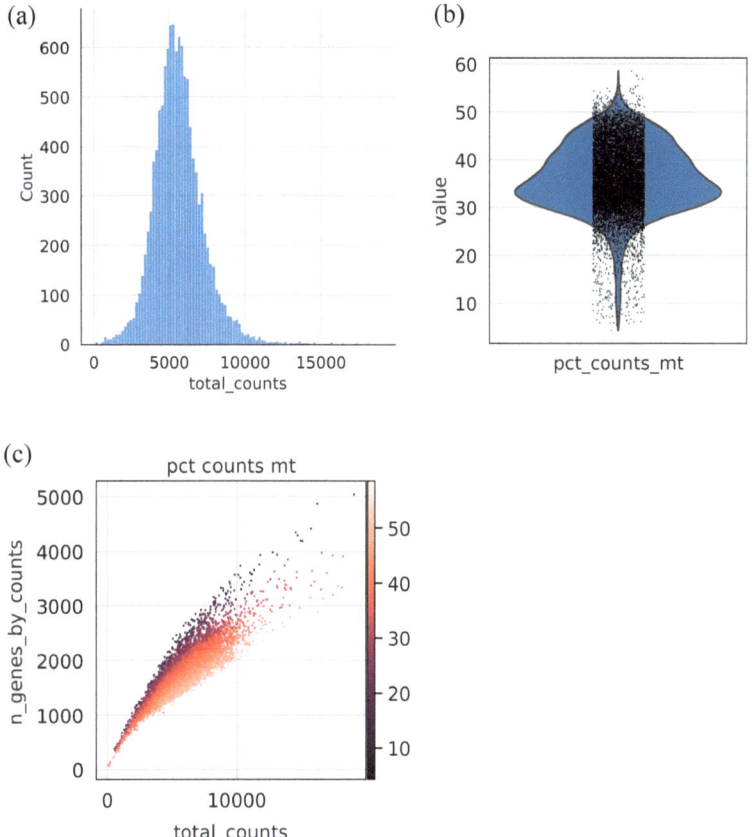

Fig. 2.3 Illustration of the QC metrics. **a** total_counts, **b** pct_counts_mt per sample, and **c** n_genes_by_counts

Fig. 2.4 Plot after applying filters

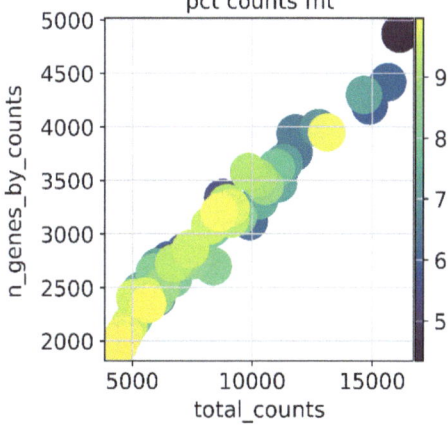

Fig. 2.5 Principal
component analysis of the
datasets

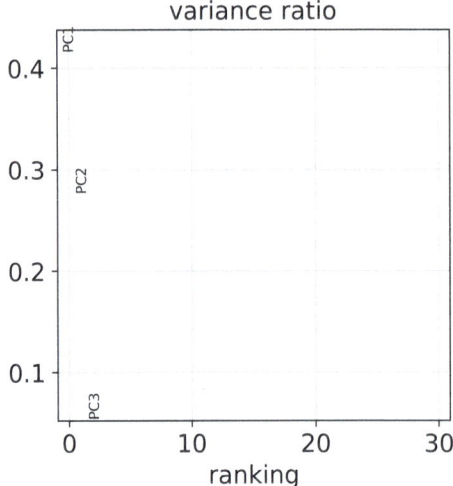

```
sc.pl.pca_variance_ratio(adata)
```

Clustering: You perform clustering to identify cell populations based on gene
expression profiles (Fig. 2.6).

```
# Clustering cells
import igraph
# UMAP is based on the neighbor graph; first compute a
neighborhood graph of observations.
sc.pp.neighbors(adata)
```

Fig. 2.6 Clustering analysis
of the datasets using UMAP

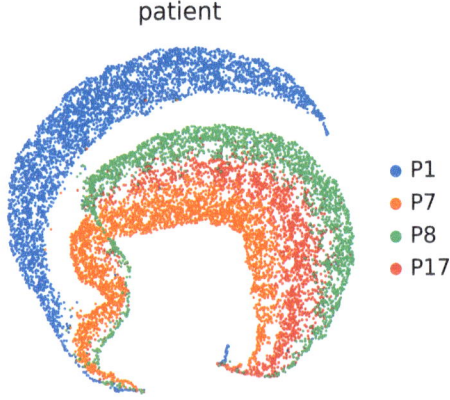

```
# add X_umap in obsm
sc.tl.umap(adata,min_dist=0.5,spread=1.0,random_state=1,n_
components=2)
sc.pl.umap(adata, color='patient')
```

2.6 Conclusion

Single-cell RNA sequencing (scRNA-seq) is a powerful technology that allows researchers to profile gene expression at the individual cell level, however, their analysis is complex and requires careful preprocessing and quality control (QC) to ensure the accuracy and reliability of downstream analyses. In this chapter, we systematically presented data preprocessing and QC metrics and their importance in scRNA-seq data analysis. We also practically demonstrated these data preprocessing and QC metrics with python code examples.

References

1. C. Ahlmann-Eltze, W. Huber, Comparison of transformations for single-cell RNA-seq data. Nat. Methods 1–8 (2023)
2. E. Becht, L. McInnes, J. Healy, C.A. Dutertre, I.W. Kwok, L.G. Ng, E.W. Newell et al., Dimensionality reduction for visualizing single-cell data using UMAP. Nat. Biotechnol. 37(1), 38–44 (2019)
3. N.J. Bernstein, N.L. Fong, I. Lam, M.A. Roy, D.G. Hendrickson, D.R. Kelley, Solo: doublet identification in single-cell RNA-seq via semi-supervised deep learning. Cell Syst. 11(1), 95–101 (2020)
4. T. Chari, L. Pachter, The specious art of single-cell genomics. PLoS Comput. Biol. 19(8), e1011288 (2023)
5. G. Chen, B. Ning, T. Shi, Single-cell RNA-seq technologies and related computational data analysis. Front. Genet. 10, 317 (2019)
6. S.J. Fleming, M.D. Chaffin, A. Arduini, A.D. Akkad, E. Banks, J.C. Marioni, M. Babadi et al., Unsupervised removal of systematic background noise from droplet-based single-cell experiments using Cell Bender. Nat. Methods 20(9), 1323–1335 (2023)
7. L. Haghverdi, F. Buettner, F.J. Theis, Diffusion maps for high-dimensional single-cell analysis of differentiation data. Bioinformatics 31(18), 2989–2998 (2015)
8. L. Haghverdi, A.T. Lun, M.D. Morgan, J.C. Marioni, Batch effects in single-cell RNA-sequencing data are corrected by matching mutual nearest neighbors. Nat. Biotechnol. 36(5), 421–427 (2018)
9. J. Hao, W. Cao, J. Huang, X. Zou, Z.G. Han, Optimal gene filtering for single-cell data (OGFSC)—a gene filtering algorithm for single-cell RNA-seq data. Bioinformatics 35(15), 2602–2609 (2019)
10. A. Haque, J. Engel, S.A. Teichmann, T. Lönnberg, A practical guide to single-cell RNA-sequencing for biomedical research and clinical applications. Genome Med. 9(1), 1–12 (2017)
11. J. Hastings, D. Lee, M.J. O'Connell, Batch-effect correction in single-cell RNA sequencing data using JIVE. bioRxiv, 2023-10 (2023)

12. L. Heumos,, A.C. Schaar, C. Lance, A. Litinetskaya, F. Drost, L. Zappia, F.J. Theis, et al., Best practices for single-cell analysis across modalities. Nat. Rev. Genet. 1–23 (2023)
13. R. Hong, Y. Koga, S. Bandyadka, A. Leshchyk, Y. Wang, V. Akavoor, J.D. Campbell et al., Comprehensive generation, visualization, and reporting of quality control metrics for single-cell RNA sequencing data. Nat. Commun. **13**(1), 1688 (2022)
14. J.A. Johnson, A.P. Tsang, J.T. Mitchell, D.L. Zhou, J., Bowden, E. Davis-Marcisak, G.L. Stein-O'Brien et al., Inferring cellular and molecular processes in single-cell data with non-negative matrix factorization using Python, R and GenePattern notebook implementations of CoGAPS. Nat. Protoc. 1–42 (2023)
15. D. Kobak, P. Berens, The art of using t-SNE for single-cell transcriptomics. Nat. Commun. **10**(1), 5416 (2019)
16. I. Korsunsky, N. Millard, J. Fan, K. Slowikowski, F. Zhang, K. Wei, S. Raychaudhuri et al., Fast, sensitive and accurate integration of single-cell data with Harmony. Nat. Methods **16**(12), 1289–1296 (2019)
17. C. Kuppe, R.O. Ramirez Flores, Z. Li et al., Spatial multi-omic map of human myocardial infarction. Nature **608**, 766–777 (2022)
18. M.D. Luecken, F.J. Theis, Current best practices in single-cell RNA-seq analysis: a tutorial. Mol. Syst. Biol. **15**(6), e8746 (2019)
19. A.T. Lun, D.J. McCarthy, J.C. Marioni, A step-by-step workflow for low-level analysis of single-cell RNA-seq data with bioconductor. F1000 Res 5 (2016)
20. N. Lytal, D. Ran, L. An, Normalization methods on single-cell RNA-seq data: an empirical survey. Front. Genet. **11**, 41 (2020)
21. D.J. McCarthy, K.R. Campbell, A.T.L. Lun, Q.F. Willis, Scater: pre-processing, quality control, normalisation and visualisation of single-cell RNA-seq data in R. Bioinformatics **33**, 1179–1186 (2017)
22. C.S. McGinnis, L.M. Murrow, Z.J. Gartner, DoubletFinder: doublet detection in single-cell RNA sequencing data using artificial nearest neighbors. Cell Syst. **8**(4), 329–337 (2019)
23. S. Rizzetto, A.A. Eltahla, P. Lin, R. Bull, A.R. Lloyd, J.W. Ho, F. Luciani et al., Impact of sequencing depth and read length on single cell RNA sequencing data of T cells. Sci. Rep. **7**(1), 12781 (2017)
24. Single-Cell Best Practices, Kuppe visium human heart 2022 control. figshare. Dataset (2023). https://doi.org/10.6084/m9.figshare.22132958.v1
25. A. Vandenbon, Evaluation of critical data processing steps for reliable prediction of gene co-expression from large collections of RNA-seq data. PLoS ONE **17**(1), e0263344 (2022)
26. W. Wang, H. Tan, M. Sun, Y. Han, W. Chen, S. Qiu, T. Ni et al., Independent component analysis based gene co-expression network inference (ICAnet) to decipher functional modules for better single-cell clustering and batch integration. Nucl. Acids Res. **49**(9), e54–e54 (2021)
27. L.L. Weber, P. Sashittal, M. El-Kebir, DoubletD: detecting doublets in single-cell DNA sequencing data. Bioinformatics **37**(Suppl. 1), i214–i221 (2021)
28. S.L. Wolock, R. Lopez, A.M. Klein, Scrublet: computational identification of cell doublets in single-cell transcriptomic data. Cell Syst. **8**(4), 281–291 (2019)
29. R. Xiang, W. Wang, L. Yang, S. Wang, C. Xu, X. Chen, A comparison for dimensionality reduction methods of single-cell RNA-seq data. Front. Genet. **12**, 646936 (2021)
30. H. Zhang, M. Lu, G. Lin, L. Zheng, W. Zhang, Z. Xu, F. Zhu, SoCube: an innovative end-to-end doublet detection algorithm for analyzing scRNA-seq data. Brief. Bioinform. **24**(3), bbad104 (2023)
31. Y. Zhang, G. Parmigiani, W.E. Johnson, ComBat-seq: batch effect adjustment for RNA-seq count data. NAR Genom. Bioinf. **2**(3), lqaa078 (2020)

Chapter 3
Dimensionality Reduction and Clustering

3.1 Background

Dimensionality reduction and clustering are two important approaches that help researchers and data analysts make sense of the high-dimensional data generated by single-cell sequencing. Dimensionality reduction helps simplify high-dimensional single-cell RNA-seq (scRNA) data for visualization and analysis, while clustering is used to identify and categorize similar cells into distinct groups, revealing the underlying structure and heterogeneity in the data. Mostly, dimensionality reduction and clustering methods are used together in scRNA data analysis pipelines in order to gain insights into the cellular composition and heterogeneity of biological samples, including disease studies [9].

Dimensionality reduction is a computational technique used to reduce the number of variables (e.g., genes or features) in a dataset while preserving the essential information [13]. In scRNA data analysis, it is often used to simplify complex data for visualization, interpretation, and downstream analysis. Examples of dimensionality reduction method are Principal Component Analysis (PCA), Uniform Manifold Approximation and Projection (UMAP), and t-distributed Stochastic Neighbor Embedding (t-SNE), discussed in this chapter. Imagine you have a scRNA dataset with thousands of genes as features. Applying PCA to this data will transform it into a smaller set of uncorrelated variables, called principal components. Each principal component is a linear combination of the original genes (or features), capturing the most significant sources of variance in the data. By reducing the dimensionality, you can visualize the data in two or three dimensions, making it easier to identify clusters and patterns among single cells [3].

On the other hand, clustering is the process of grouping similar data points together based on metrics such as Euclidean distance, Manhattan distance, Cosine similarity, and so on [10]. In scRNA data analysis, clustering is applied to identify distinct cell populations or cell types within a heterogeneous data sample. These clusters represent groups of cells with similar gene expression profiles. Example of clustering

© The Author(s), under exclusive license to Springer Nature Singapore Pte Ltd. 2024   31
K. Raza, *Machine Learning in Single-Cell RNA-seq Data Analysis*,
SpringerBriefs in Computational Intelligence,
https://doi.org/10.1007/978-981-97-6703-8_3

method is k-means. Let's say you have a scRNA dataset with the expression profiles of hundreds of cells. You may apply k-means clustering to group these cells into k clusters (where k is a user-defined parameter). The algorithm assigns each cell to the cluster that minimizes the distance to the cluster's center. After clustering, you can label each cluster as a particular cell type or state based on the gene expression patterns within each cluster [10].

3.2 PCA for Dimensionality Reduction

Principal Component Analysis (PCA) is a widely accepted technique for dimensionality reduction in scRNA-seq data analysis. This technique helps reduce the high dimensionality of scRNA data while preserving the most critical information and patterns. Here's an explanation of how PCA works, along with an example and some software tools commonly used for PCA in scRNA-seq analysis.

3.2.1 How PCA Works?

PCA is a linear transformation method that identifies and ranks the principal components (PCs) in a dataset. These PCs are linear combinations of the original features (genes in the case of scRNA data) and are orthogonal to each other, meaning they are uncorrelated. The first PC (PC_1) captures the maximum variance in the data, the second PC (PC_2) captures the second most significant variance, and so on. By selecting a subset of these PCs, you can effectively reduce the dimensionality of the data [5]. Since variations can be capture by analyzing the correlated behavior of several genes, the top PCs are likely to present the biological signal and dominant features of heterogeneity are captured by top PCs. Hence, restricting our analysis to top PCs will then reduce the dimension of the data.

3.2.2 PCA Algorithm

The PCA algorithm comprises of the following major steps:

Data preprocessing: Start by normalizing and scaling scRNA data to ensure that all genes have the same variance, i.e., give equal importance to all genes.

Covariance matrix calculation: This step aims to see if there is any relationship between variables. If variables are highly correlated, it means that they contain redundant information. Hence, covariance matrix is computed to quantify relationships between genes [4].

Eigenvalue decomposition: Perform eigenvalue decomposition on the covariance matrix to find the eigenvalues and eigenvectors. The eigenvectors represent the PCs, and the eigenvalues represent the variance explained by each component.

- *Select PCs*: Sort the eigenvalues in descending order and select the top "k" eigenvectors to reduce the dimensionality to "k" dimensions.
- *Transform data*: Multiply the original data by the selected eigenvectors to obtain the reduced-dimension representation.

Example Suppose you have an scRNA dataset with the expression levels of 10,000 genes across 1000 cells. Applying PCA with the goal of reducing the data to two dimensions, you obtain two principal components, PC1 and PC2. These PCs capture the most significant patterns in the data, such as the primary sources of variation among the cells.

3.2.3 Tools and Software Tools for PCA in scRNA-seq Data Analysis

A plethora of software tools for PCA are available. Some of the popular used ones are described as follows:

- *R and Bioconductor*: The R programming language provides numerous packages within the Bioconductor project (https://bioconductor.org/), such as Seurat v5 integration or Harmony [6, 7], and SingleR [2], which offer PCA functionality and integration with other scRNA-seq analysis tools.
- *Python*: Python libraries like scikit-learn, numpy, and PCA from SciPy are commonly used for PCA in scRNA-seq data analysis. Packages like Scanpy [12] is also popular choice.
- *MATLAB*: MATLAB offers built-in functions for PCA and data analysis, making it suitable for those comfortable with MATLAB programming.
- *Single-cell analysis software*: Specialized single-cell RNA-seq analysis tools like Seurat, Scanpy, and Monocle often include PCA as a component of their workflows.

3.2.4 Case Study

Here's a demonstration of how to apply PCA for dimensionality reduction in scRNA-seq data using Python and Scanpy. In this example, we'll assume that you have already installed Scanpy and loaded your scRNA-seq data into an Anndata (a fundamental data structure) object.

A sample code block showing applications of PCA using Scanpy:

```python
# Import the necessary libraries
import scanpy as sc
# Load your scRNA-seq data into an Anndata object
# Assuming that file is not your local drive
adata = sc.read("your_scRNA_data.h5ad")
# Preprocess your data (if necessary)
# You may perform QC, filtering, and normalization
# Run PCA, applied PCA on adata
# Adds results in .obsm, .varm, .uns
sc.pp.pca(adata)
# Visualize PCA results
sc.pl.pca_overview(adata,      color="interesting_variable",
save="_pca.pdf")
# Additional analyses can be performed using the
reduced-dimension data
# You can proceed with clustering, differential expression
analysis, etc.
# Save the processed data for future analysis
adata.write("processed_scRNA_data.h5ad")
```

Explanation of the above code block:

We import Scanpy and load scRNA-seq data into an Anndata object (adata). Be sure to replace "your_scRNA_data.h5ad" with the actual file path of your data. If necessary, perform data preprocessing steps like QC, filtering, and normalization before running PCA, which is important for ensuring the quality of analysis. In the above code block, we run PCA on the data using **sc.pp.pca()** function. This function computes PCA coordinates, loadings and variance decomposition, and results are stored in .obsm, .varm, .uns attributes of AnnData object adata. There are several parameters that may be passed to this function such as number of principal components (n_comps, default to 50), SVD solver to use (svd_solver, default to 'auto', however it can be 'arpack', 'randomized', etc.).

One of the purposes of dimensionality reduction using PCA is to compress data for plotting into two or three dimension with the most influential features of data. To visualize the PCA results, we have used **sc.pl.pca_overview()** function which generates a summary plot providing an overview of the PCA results. You can color the plot based on variables of interest. The output of **sc.pl.pca_overview()** typically includes (i) Scree plot, (ii) PC heatmap, (iii) PC scatter plot, and (iv) Optional annotations. A scree plot depicts the variance ratio of each PC, which helps us in determining the number of PCs to retain for downstream analysis by visualizing the proportion of variance explained by each component. A heatmap shows the loadings of each variable (e.g., genes) on the PCs which provides insights into the relationships between the original variables and the PCs and helps in identifying genes that contribute most to the variance captured by each component. A scatter plot visualizes the samples (cells) in the reduced dimensional space defined by the first few PCs which helps in visualizing the clustering or separation of samples based on their PCA coordinates and understanding the underlying structure of the data. Further, additional annotations such as cell type labels or experimental conditions can be overlaid on the scatter plot to provide further insights into the biological or experimental factors driving the observed variability. After the dimensionality reduction, you can proceed with additional analyses, such as clustering and differential expression, using the reduced-dimension data. Finally, we saved the processed data for future analysis.

In order to make the reader understand the output of PCA, we have executed the following Python code on Google Colab Engine (https://colab.research.google.com/). In this code example, we directly loaded the scRNA-seq dataset from a publicly available repository using its URL (Kuppe Visium Human Heart 2022 Control datasets, already explained in Chap. 2).

```
# Important required packages
import scanpy as sc
import requests
from io import BytesIO
# Load the scRNA-seq data from a URL
# Source: https://doi.org/10.6084/m9.figshare.22132958.v1
url = 'https://figshare.com/ndownloader/files/39347357'
response = requests.get(url) # Download the .h5ad file
response.raise_for_status()  # Check if the request was
successful
# Read the downloaded content into an AnnData object
adata = sc.read_h5ad(BytesIO(response.content))
adata.var_names_make_unique()
# Basic filtering: filter out genes detected in less than
3 cells
sc.pp.filter_cells(adata, min_genes=300)
sc.pp.filter_genes(adata, min_cells=3)
# Visualize top 20 genes having highest counts in each
single cell
sc.pl.highest_expr_genes(adata, n_top=20)
# Applying PCA with default parameters:
# Adds PCA results in .obsm, .varm, .uns
sc.pp.pca(adata)
# Plots PCA results with color as MT-CO1 variable
sc.pl.pca_overview(adata, color='MT-CO1')
# Saving results for future use
adata.write("processed_scRNA_data.h5ad")
```

The output of the code generates few figures. The results of top 20 genes having highest counts in each single cell has been depicted as boxplot in Fig. 3.1. Further, Fig. 3.2 shows that the first principal component (PC_1) which captures the majority of the variation in the data, and that a single gene dominates the variation in that component. In order to inspect the contribution of single PCs to the total variance in the data, we depicted elbow plot shown in Fig. 3.3, which provides information about how many PCs (known as elbow point) we should consider to compute the neighborhood relations of cells. To find the elbow point, we should choose a last PC from which percent of variance should not drop. From Fig. 3.3 PC3 can be observed as elbow point, stating that total number of PCs can be considered as 3. Figure 3.4 depicts the loading, i.e., how strongly each gene contributes to principle components PC1, PC2 and PC3. We can observe that these PCs are determined by the gene expression of just a small number of genes.

Fig. 3.1 Boxplot of the top 20 genes having highest counts in each single cell

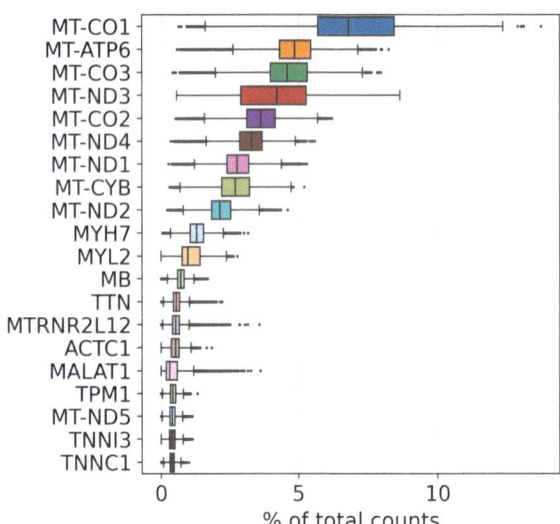

Fig. 3.2 Scatter plot of the PCA coordinates for gene MT-CO1

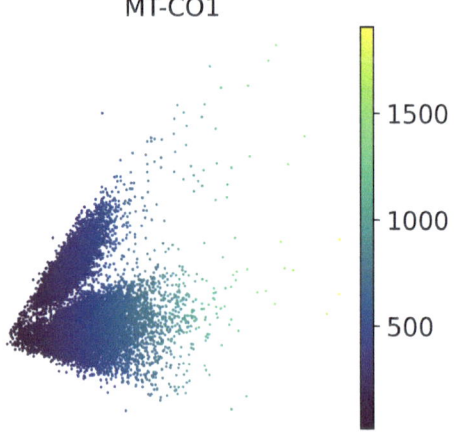

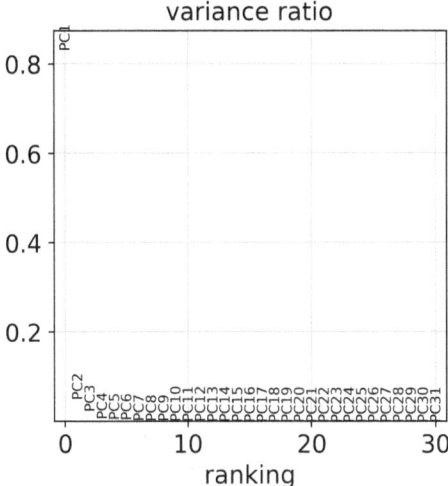

Fig. 3.3 Elbow plot showing the rank of principal components (PCs) based on percent of variance

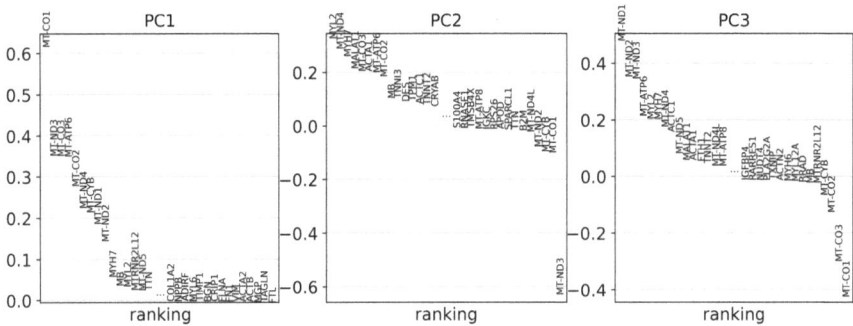

Fig. 3.4 Plots showing the load, indicating how strongly each gene contributes to each principle components (PCs) such as PC1, PC2 and PC3

3.3 t-SNE

t-SNE (t-distributed Stochastic Neighbor Embedding) is a powerful technique for visualizing high-dimensional data, including scRNA-seq. It is commonly used in scRNA-seq data analysis to reduce the dimensionality of the data and create a two-dimensional (2D) or three-dimensional (3D) representation of the cells, allowing researchers to explore and visualize the underlying structure of the data [11].

3.3.1 How t-SNE Works?

t-SNE is a nonlinear dimensionality reduction technique that emphasizes preserving local similarities in the high-dimensional space. It achieves this by modeling the pairwise similarity between data points in both the high-dimensional and low-dimensional spaces. In the low-dimensional space, t-SNE tries to represent similar data points as close neighbors and dissimilar data points as more distant (Van der Maaten and Hinton 2008).

3.3.2 t-SNE Algorithm

Compute Pairwise Similarities: Calculate pairwise similarities between data points in the high-dimensional space. In scRNA-seq data analysis, these data points represent individual cells, and the similarity is typically calculated based on gene expression profiles (e.g., Euclidean distance or other similarity measures).

Construct Conditional Probability Distributions: Create conditional probability distributions for pairwise similarities in both the high-dimensional and low-dimensional spaces. This step involves transforming the similarities into probability distributions. t-SNE uses a Gaussian kernel to convert pairwise similarities into probabilities.

Optimize Low-Dimensional Representation: Use an optimization algorithm (typically gradient descent) to find a low-dimensional representation where the pairwise similarities match as closely as possible to the high-dimensional space. The optimization minimizes the Kullback–Leibler divergence between the conditional probability distributions in the high-dimensional and low-dimensional spaces.

Reduce Dimensionality: The output of the t-SNE algorithm is a lower-dimensional representation of the data (often 2D or 3D) where similar data points are close to each other, and dissimilar data points are far apart.

Example Suppose you have an scRNA-seq dataset with gene expression profiles for 1000 cells. You can use t-SNE to reduce the dimensionality of this data while preserving the inherent similarities and differences between cells. The output will be a 2D or 3D representation of the cells, where clusters of similar cells are visually apparent.

3.3.3 Tools and Software for t-SNE in scRNA-seq Data Analysis

Scikit-learn: Scikit-learn is a widely used Python library that includes a t-SNE implementation.

Seurat: Seurat, an R package for single-cell data analysis, includes t-SNE as a part of its workflow.

SCANPY: SCANPY, a Python library for single-cell RNA-seq analysis, provides functions to run t-SNE on scRNA-seq data. It's particularly useful for end-to-end scRNA-seq analysis pipelines.

Rtsne: This is an R package specifically designed for t-SNE dimensionality reduction. It provides an easy-to-use implementation of t-SNE in R.

3.3.4 Case Study

Here's a demonstration of how to apply t-SNE for dimensionality reduction in scRNA-seq data using Python and Scanpy. We'll assume you have already loaded your scRNA-seq data into an Anndata object 'adata'.

```
# Run t-SNE
sc.tl.tsne(adata,    n_pcs=20,    perplexity=30,    learning_
rate=300)
# Visualize the t-SNE results
sc.pl.tsne(adata, color=['MT-CO1', 'MYH7'])
sc.pl.tsne(adata, color="patient")
sc.pl.tsne(adata, color="batch")
sc.pl.tsne(adata, color="patient_region_id")
```

Explanation
Import Scanpy and load your scRNA-seq data into an Anndata object (adata). If needed, you can perform data preprocessing. Run t-SNE on the data using `sc.tl.tsne()` function. You can specify the number of principal components (n_pcs), perplexity, and learning rate based on your data and analysis needs. The perplexity parameter (usually defaults to 30) is considered to have the balance between preserving the global and the local structure of the data, i.e., it is the contin-uous analogy to the k number of nearest neighbors for which distances will be preserved. The learning rate parameter (usually ranges between 10.0 and 1000.0) controls the step size of the gradient updates. Visualize the t-SNE results using `sc.pl.tsne()`. You can color the t-SNE plot based on variables of interest, such

as cell type, gene expression levels, or any metadata associated with your cells. Finally, save the processed data, including the t-SNE results, for future analysis.

We executed the above t-SNE code on the same scRNA-seq data considered in the previous case study. The obtained results are depicted in Fig. 3.5.

Figure 3.5a depicts MT-CO1 and MYH7 across clusters, showing that both of these genes are strongly expressed in cluster 0, with different count ranges. In Fig. 3.5b, the cluster numbering is based on the size, i.e., clusters 0 and cluster 1 are not essentially related, they are just clusters containing the most cells in patient and batch. In general, PCA is not sufficient to analysis clusters in samples, however, t-SNE visualization are useful.

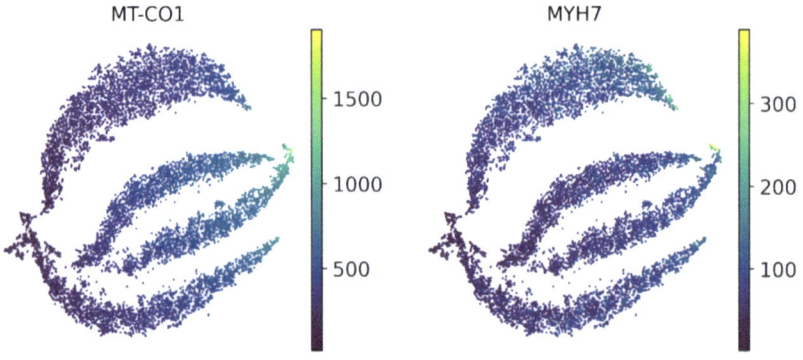

(a)   MT-CO1 and MYH7 across clusters

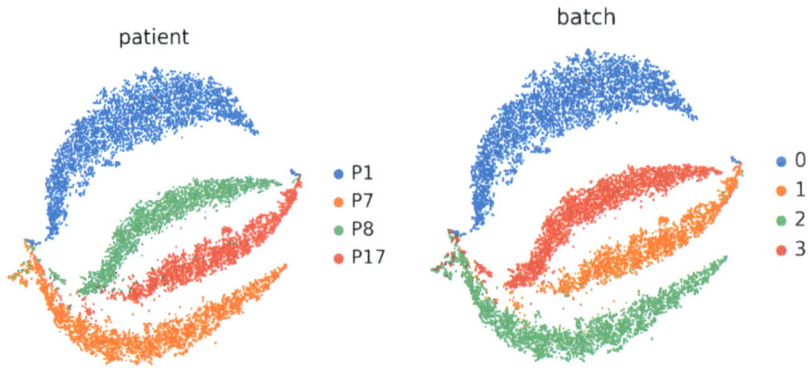

(b) Patient and batch clusters

Fig. 3.5 The results of t-SNE analysis

3.4 Clustering Algorithms

Clustering is an unsupervised learning method used to identify groups of cells having similar expression profiles within scRNA-seq samples. It simplifies complex genetic data into discrete, interpretable labels, helping to describe population heterogeneity. Once clusters are annotated with marker genes, they can represent biological concepts like cell types or states, making the data more accessible for analysis. Each cluster represents a putative cell population with shared gene expression patterns. This allows researchers to identify and categorize different cell types, subtypes, or states within a heterogeneous sample. Similarly, clustering techniques also enable the discovery of novel cell types or subpopulations by revealing hidden patterns in the data. By examining the marker genes and biological characteristics of each cluster, researchers can identify and characterize previously unknown cell types [8].

Common clustering techniques used in scRNA-seq data analysis include *k-means*, *hierarchical clustering*, *DBSCAN* (Density-Based Spatial Clustering of Applications with Noise), and graph-based approaches like *Louvain clustering* or *Seurat's graph-based clustering*. The choice of clustering method depends on the characteristics of the data, the complexity of the cellular heterogeneity, and the research goals.

3.4.1 K-Means Clustering

The k-means is a widely used clustering method that can be applied to scRNA-seq data to identify distinct cell populations or clusters based on gene expression profiles. Before applying k-means, we need to preprocess scRNA-seq data, as already discussed earlier. Due to high dimensionality of data, you need to choose relevant features (genes) that capture the most variance in the data. This step reduces noise and computational complexity. Apply k-means clustering to group cells with similar gene expression profiles into clusters. k-means partitions data into k clusters, where k is a user-defined parameter representing the desired number of clusters. After that you need to assess the quality of the clustering results using metrics like silhouette score, Davies-Bouldin index, or visual inspection. Finally, analyze and interpret the biological significance of the identified cell clusters. This may involve looking at marker genes and functional enrichment analysis [14].

K-means clustering for analyzing scRNA-seq data is available in several packages and platforms, each offering various tools for preprocessing, clustering, and visualization. Some of the popular ones are Scanpy (integrates with scikit-learn for k-means clustering), Seurat (direct support for k-means clustering, Cumulus (a cloud-based platform for large-scale scRNA-seq data analysis developed by the Broad Institute), Cell Ranger (A software pipeline provided by 10× Genomics, primarily offering graph-based clustering methods, it can be used in conjunction with downstream analysis in R or Python that supports k-means), Monocle (provides clustering options including k-means through integration with other R functions).

Here's a demonstration of how to apply k-means in scRNA-seq data using Python and Scanpy. We'll assume you have already loaded your scRNA-seq data into an Anndata object 'adata'.

```
# Import necessary packages
import scanpy as sc
from sklearn.cluster import KMeans
# Apply PCA
sc.pp.pca(adata)
# Perform k-means clustering
kmeans          =          KMeans(n_clusters=10,          random_
state=0).fit(adata.obsm['X_pca'])
# Add cluster labels to the AnnData object
adata.obs['kmeans'] = kmeans.labels.astype(str)
# Compute UMAP for visualization
sc.pp.neighbors(adata, n_neighbors=5, n_pcs=10)
sc.tl.umap(adata)
# Visualize the identified clusters
sc.pl.umap(adata,   color='kmeans',   title="K-means   Clus-
tering")
```

Explanation

Import scanpy, sklearn, and load your scRNA-seq data into an Anndata object (adata). If needed, you can perform data preprocessing. Run PCA on the data using `sc.pp.pca()` function for dimensionality reduction. Fit the k-means algorithm on the PCA-reduced data using `KMeans()` function and add the resulting cluster labels to the AnnData object. Compute the UMAP embedding for visualization, and plot the UMAP, coloring the cells by the k-means cluster labels. The code produces a UMAP plot where cells are colored according to their k-means cluster assignments, allowing you to visualize the clustering results. Adjust the number of clusters (n_ clusters) based on your specific dataset and analysis needs.

We executed the above k-means code on the same scRNA-seq data considered in the previous case study. The obtained clustering result is shown in Fig. 3.6, presenting cells clustered in k = 10 different clusters.

3.4.2 Graph-Based Clustering

Graph-based clustering is a method where the data is represented as a graph, with nodes representing data points (cells, in the case of scRNA-seq) and edges representing the similarity or distance between them. The graph structure is then used to identify clusters of closely connected nodes, often using algorithms like Louvain or Leiden. Graph-based clustering is extensively used in scRNA-seq data analysis due to its capability to handle the high-dimensionality and sparsity inherent in the data.

Fig. 3.6 Cells are clustered (grouped) using k-means clustering with k = 10

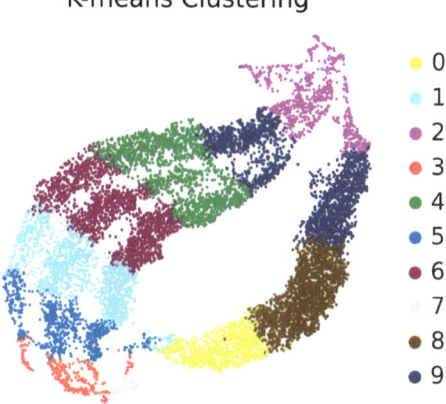

Like other clustering algorithms, graph-based clustering can be used for grouping cells with similar expression profiles to infer distinct cell types, discovering different states or stages of cells (e.g. various differentiation stages), and exploring the diversity within a cell population to understand complex biological processes. The graph-based clustering is usually preferred over k-means and other algorithms because of its flexibility in cluster shape, scalability, resolution parameter, and robustness to noise [1].

The most commonly used graph-based clustering methods are Louvain Method (designed to optimize the modularity of a partition of a network), Leiden Algorithm (improved version of the Louvain method that guarantees well-connected communities and is faster and more robust), Walktrap (uses random walks on the graph to detect communities), Label Propagation (a simple and quick, used for large-scale graph clustering), and so on.

Here's a demonstration of how to apply Louvain graph-based method in scRNA-seq data using Python and Scanpy. We'll assume you have already loaded your scRNA-seq data into an Anndata object 'adata'. You may need to install some additional packages, like pip install igraph, and leidenalg. Run command **pip install scanpy igraph leidenalg**.

```python
import scanpy as sc
# Compute the neighborhood graph
sc.pp.neighbors(adata, n_neighbors=10, n_pcs=50)
# Perform graph-based clustering using Leiden algorithm
sc.tl.leiden(adata, resolution=0.5)
# Compute UMAP for visualization
sc.tl.umap(adata)
# Visualize the identified clusters
sc.pl.umap(adata, color='leiden', title="Leiden Clustering
(Graph-based)")
```

Fig. 3.7 Graph-based
Leiden clustering of cells

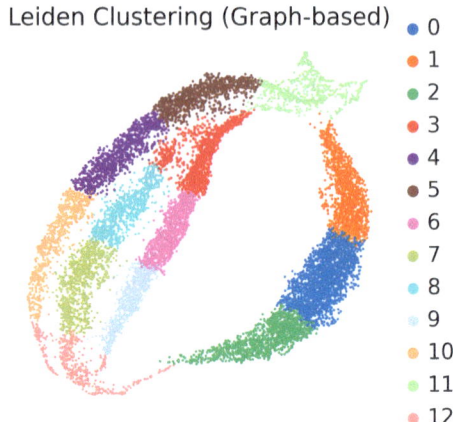

Explanation of Code

We assume that you have already loaded your data to AnnData object 'adata, performed data preprocessing steps, and dimensionality reduction using PCA or t-SNE. For performing graph-based clustering, we first need to compute the Neighborhood Graph using a k-nearest neighbors graph based on the PCA-reduced data. For this, we used `sc.pp.neighbors()` function of Scanpy. We assumed k = 10 (`n_ neighbors` parameter), and `n_pcs=50`, what can be set as per requirements. After that, we may perform graph-based clustering using the Leiden algorithm on the neighborhood graph to identify clusters. For this, we invoked `sc.tl.leiden()` function with parameter resolution = 0.5. Resolution parameter value controls the coarseness of the clustering. Higher values lead to more clusters. In order to apply Louvain method, we can use function `sc.tl.louvain()`. Once clusters are identified, we need to compute the UMAP embedding for visualization using `sc.tl.umap()` function. Finally, we can plot the identified clusters using sc.pl.umap(), coloring the cells by the Leiden cluster labels. The output of the above code is depicted in Fig. 3.7, showing clusters of cells.

3.5 Conclusion

Both dimensionality reduction and clustering are two important tasks that we perform for the analysis of scRNA-seq data. This chapter systematically presents dimensionality reduction approaches such as PCA and t-SNE, along with its implementation using Python and Scanpy. Further, clustering methods such as k-means and graph-based (Leiden algorithm) clustering approaches, and their implementations have also been discussed.

References

1. S.A.R. Abadi, S.P. Laghaee, S. Koohi, An optimized graph-based structure for single-cell RNA-seq cell-type classification based on non-linear dimension reduction. BMC Genom. **24**(1), 227 (2023)
2. D. Aran, A.P. Looney, L. Liu, E. Wu, V. Fong, A. Hsu, S. Chak, R.P. Naikawadi, P.J. Wolters, A.R. Abate, A.J. Butte, M. Bhattacharya, Reference-based analysis of lung single-cell sequencing reveals a transitional profibrotic macrophage. Nat. Immunol. **20**, 163–172 (2019). https://doi.org/10.1038/s41590-018-0276-y
3. C. Feng, S. Liu, H. Zhang, R. Guan, D. Li, F. Zhou, X. Feng et al., Dimension reduction and clustering models for single-cell RNA sequencing data: a comparative study. Int. J. Mol. Sci. **21**(6), 21–81 (2020)
4. D. Groth, S. Hartmann, S. Klie, J. Selbig, Principal components analysis. Comput. Toxicol. **II**, 527–547 (2013)
5. I.T. Jolliffe, J. Cadima, Principal component analysis: a review and recent developments. Philos. Trans. R. Soc. A Math. Phys. Eng. Sci. **374**(2065), 20150202 (2016)
6. I. Korsunsky, N. Millard, J. Fan, K. Slowikowski, F. Zhang, K. Wei, S. Raychaudhuri et al., Fast, sensitive and accurate integration of single-cell data with Harmony. Nat. Methods **16**(12), 1289–1296 (2019)
7. A.T. Lun, D.J. McCarthy, J.C. Marioni, A step-by-step workflow for low-level analysis of single-cell RNA-seq data with bioconductor. F1000 Res. 5 (2016)
8. L. Peng, X. Tian, G. Tian, J. Xu, X. Huang, Y. Weng, L. Zhou et al., Single-cell RNA-seq clustering: datasets, models, and algorithms. RNA Boil. **17**(6), 765–783 (2020)
9. S. Sun, J. Zhu, Y. Ma, X. Zhou, Accuracy, robustness and scalability of dimensionality reduction methods for single-cell RNA-seq analysis. Genome Biol. **20**, 1–21 (2019)
10. R. Suwanda, Z. Syahputra, E.M. Zamzami, Analysis of euclidean distance and manhattan distance in the K-means algorithm for variations number of centroid K. J. Phys. Conf. Ser. **1566**(1), 012058 (2020)
11. L. Van der Maaten, G. Hinton, Visualizing data using t-SNE. J. Mach. Learn. Res. 1; **9**(11), (2008)
12. F.A. Wolf, P. Angerer, F.J. Theis, SCANPY: large-scale single-cell gene expression data analysis. Genome Biol. **19**, 1–5 (2018)
13. R. Xiang, W. Wang, L. Yang, S. Wang, C. Xu, X. Chen, A comparison for dimensionality reduction methods of single-cell RNA-seq data. Front. Genet. **12**, 646936 (2021)
14. X. Zhu, H.D. Li, L. Guo, F.X. Wu, J. Wang, Analysis of single-cell RNA-seq data by clustering approaches. Curr. Bioinform. **14**(4), 314–322 (2019)

Chapter 4
Differential Expression Analysis

4.1 Background and Motivation

In the rapidly evolving landscape of genomics, the quest to unravel the intricacies of gene expression has undergone a transformative shift. Traditional methods, such as bulk RNA-seq, have proven invaluable in providing a snapshot of cellular activity. However, these approaches inherently mask the rich heterogeneity inherent in complex biological systems. The emergence of single-cell RNA sequencing (scRNA-seq) has, therefore, become a pivotal milestone in the pursuit of a more nuanced understanding of gene expression dynamics. Traditional bulk RNA-seq, while groundbreaking, lacks the resolution to decipher the nuanced gene expression patterns within heterogeneous cell populations. The need for a more granular approach became apparent as researchers sought to untangle the intricate tapestry of cellular diversity [10].

Gene expression is the process by which a gene's genetic code is utilized to manage the synthesis of proteins necessary for building cell structures in the body. Differential gene expression refers to the process where different genes are activated within a cell, giving the cell a specific purpose and defining its function [20]. Differential expression analysis (DEA) is an RNA-seq technology used to evaluate, quantify, and compare the gene expression of cells. It identifies which genes are actively being expressed within the cell. By performing this analysis on cells from individuals with certain medical conditions, such as cancer or other diseases, we can gain insights into the causes of these conditions and potential treatments. Differentially expressed genes (DEGs) are genes that show significant differences in expression levels between different conditions, such as healthy versus diseased states. Identifying DEGs is crucial for understanding the molecular basis of diseases, discovering diagnostic biomarkers, and developing new therapeutics. For instance, in cancer research, DEGs can reveal which genes are upregulated or downregulated in tumor cells compared to normal cells, providing insights into cancer progression and potential therapeutic targets [10, 20].

© The Author(s), under exclusive license to Springer Nature Singapore Pte Ltd. 2024   47
K. Raza, *Machine Learning in Single-Cell RNA-seq Data Analysis*,
SpringerBriefs in Computational Intelligence,
https://doi.org/10.1007/978-981-97-6703-8_4

4.1.1 Motivation for Single-Cell Differential Expression Analysis

The motivation behind single-cell differential expression analysis lies in its promise to provide a more precise molecular profile. By enhancing sensitivity to cellular heterogeneity and offering improved resolution in expression profiling, this approach promises to reveal hidden nuances in gene regulation. Its ability to uncover rare cell types and states is particularly valuable, as it not only advances our fundamental understanding of biology but also holds immense potential for targeted therapeutic interventions. Further, technological advances in scRNA-seq, including droplet-based and plate-based technologies, have catalyzed the widespread adoption of single-cell approaches. The simultaneous rise in public repositories and databases, such as NCBI's Gene Expression Omnibus (GEO) and Sequence Read Archives (SRA), EBI's ArrayExpress and Single Cell Expression Atlas, Broad Institute's Single Cell Portal, Human Cell Atlas (HCA) Data Portal, has democratized access to single-cell datasets, fostering collaboration and enabling researchers to explore new frontiers in gene expression analysis.

4.1.2 Applications Across Biological Disciplines

The versatility of single-cell differential expression analysis is evident in its applications across diverse biological disciplines. From unraveling the complexities of tumor heterogeneity in cancer research to identifying neuronal subtypes in neuroscience and profiling immune responses in immunology, the impact of this methodology is far-reaching. Its implications extend beyond understanding biological systems; it holds the key to personalized medicine, treatment optimization, and innovative therapeutic strategies [5].

Scope and Structure of the Chapter

This chapter aims to provide a comprehensive guide to the principles, tools, and methodologies involved in single-cell differential expression analysis. It sets out to demystify the intricacies of this field, empowering researchers to leverage the power of single-cell data in their explorations. Through a structured exploration of various aspects, the chapter aspires to contribute to the evolving narrative of gene expression analysis, bridging the gap between theory and practical application.

4.2 Differential Expression Analysis: An Overview

Differential Expression Analysis (DEA) is a crucial method in genomics that aims to identify and quantify changes in gene expression levels between different biological conditions or experimental groups. It is particularly valuable for understanding how genes are regulated and how these regulations contribute to the observed phenotypic differences. DEA is commonly employed in studies comparing diseased and healthy states, different developmental stages, or responses to external stimuli [5].

4.2.1 Necessity of Differential Expression Analysis

Understanding Molecular Mechanisms: DEA helps researchers unravel the molecular mechanisms underlying various biological processes. By identifying genes that are differentially expressed, we gain insights into the pathways and networks involved in specific conditions.

Biomarker Discovery: DEA is instrumental in identifying potential biomarkers associated with diseases. Differential expression patterns can highlight genes that serve as indicators of disease presence, progression, or response to treatment, facilitating the development of diagnostic tools and personalized medicine [1, 17].

Drug Discovery and Target Identification: In drug development, DEA aids in the identification of potential therapeutic targets. By pinpointing genes whose expression levels are altered in disease states, we can focus on developing drugs that specifically modulate these targets, leading to more effective treatments.

Characterizing Disease Subtypes: DEA helps in delineating molecular subtypes within a disease, providing a finer understanding of its heterogeneity. This knowledge is critical for tailoring treatment strategies, as different subtypes may respond differently to therapies.

4.2.2 Role of Differential Expression Analysis in Disease Study

Apart from understanding molecular mechanisms of biological systems, biomarkers discovery, drug design and target identification, and characterizing disease subtypes, DEA has potential role in disease studies. In cancers, DEA has been employed to identify genes differentially expressed between tumor and normal tissues. This approach has led to the discovery of subtypes with distinct molecular signatures, enabling more targeted therapies. For instance, the identification of HER2-positive breast cancer through DEA has paved the way for targeted therapies like trastuzumab [18]. In an RNA-seq based DEA analysis, DLC1 has been reported as therapeutic target for specific subsets of gastric cancer [20]. In neurological disorders, such as Alzheimer's disease, DEA has been utilized to compare gene expression profiles between healthy and diseased brains. This has revealed alterations in genes related to neuroinflammation and synaptic function that provides insights into the mechanism of disease and potential therapeutic targets. In infectious diseases study, DEA can unveil host responses by identifying differentially expressed genes (DEGs) upon infection. For instance, in the study of COVID-19, DEA has been used to understand the host immune response and identify potential drug targets by analyzing gene expression changes in infected individuals compared to healthy controls [11].

Similarly, in autoimmune diseases such as rheumatoid arthritis, DEA has contributed to understanding the dysregulation of immune-related genes. Identification of DEGs in synovial tissue has led to the development of targeted therapies, such as TNF inhibitors, which have revolutionized the treatment of this autoimmune disorder [1, 12].

4.3 Statistical Methods and Tools

Several statistical methods have been developed to perform DEA on scRNA-seq data. Each method addresses the unique challenges associated with analyzing data at the single-cell level, such as sparsity, dropout events, and cell-to-cell variability. Some commonly used statistical methods are described as follows:

DESeq2: Originally designed for bulk RNA-seq, DESeq2 [16] has been adapted for scRNA-seq. It uses a negative binomial distribution to model count data and employs a shrinkage estimator to account for variability. DESeq2 provides robust estimates of gene expression changes while handling the high sparsity and dropout events characteristic of scRNA-seq data. DESeq2 works with raw count data and includes steps for data normalization, estimation of size factors, and differential expression testing. It is widely used for identifying DEGs in scRNA-seq data.

edgeR: Similar to DESeq2, edgeR (Empirical analysis of Digital Gene Expression data in R) [4] was initially designed for bulk RNA-seq but has been adapted for scRNA-seq analysis. It utilizes a negative binomial model to account for over-dispersion in count data and employs empirical Bayes methods for estimating variance and differential expression. edgeR provides a robust statistical framework for DEA in scRNA-seq. It includes steps for data normalization, estimation of dispersions, and identification of DEGs.

MAST: Model-based Analysis of Single-cell Transcriptomics (MAST) [7] is specifically designed for scRNA-seq analysis and employs a zero-inflated negative binomial model to address dropout events common in single-cell data. MAST provides accurate estimates of differential expression by accounting for both technical and biological variability. MAST includes steps for data normalization, handling zero inflation, and modeling the biological variability across single cells. It is particularly useful for detecting subtle expression changes in specific cell subpopulations.

SCDE: Single-Cell Differential Expression (SCDE) [13] is a Bayesian approach that models the count data with a negative binomial distribution and incorporates information about cell-specific expression variability. It aims to provide accurate estimates of differential expression by considering the heterogeneity among single cells. SCDE includes steps for data normalization, modeling technical and biological variability, and assessing the significance of differential expression.

BASiCS: Bayesian Analysis of Single-Cell Sequencing data (BASiCS) [19] is a Bayesian hierarchical model designed to handle both technical and biological variability in scRNA-seq data. It models the count data with a negative binomial distribution and separates biological and technical components to improve estimation accuracy. This method includes steps for data normalization, modeling technical variability, and estimating differential expression. It is particularly useful for scRNA-seq studies involving time-course experiments or other dynamic processes.

IDEAS: Individual level Differential Expression Analysis for Single cells (IDEAS) [22] is a statistical method specifically designed to analyze scRNA-seq data involving comparisons between multiple individuals. Unlike traditional scRNA-seq DEA methods that compare expression between cell groups within a single individual, IDEAS considers expression at the individual level. For each gene, IDEAS first estimates the distribution of its expression in each individual. This can be done using parametric (e.g., negative binomial) or non-parametric methods, accounting for cell-level covariates if needed. IDEAS then statistically compares these individual-specific expression distributions between the groups of interest. This helps identify genes with expression patterns that differ significantly between individuals across groups.

The above statistical methods provide robust frameworks for analyzing scRNA-seq data and identifying DEGs across individual cells. The choice of a specific method depends on the characteristics of the data and the assumptions that align with the experimental design. Researchers often consider factors such as sparsity, dropout events, and the need for accurate estimation when selecting the most appropriate statistical method for their scRNA-seq differential expression analysis. Detailed discussion on the statistical methods for DEA is beyond the scope of this Chapter, however it can be found in Das et al. [6].

4.4 Machine Learning Approaches

Machine learning approaches are increasingly playing a significant role in analyzing scRNA-seq data, including the identification of DEGs across individual cells. Some of the common deep learning approaches for DEGs are Convolutional Neural Networks (CNNs), Recurrent Neural Networks (RNNs), and Autoencoders. CNN models are adept at learning complex spatial patterns from scRNA-seq data, particularly valuable for identifying DEGs associated with specific cellular structures or subcellular localization, while RNNs excel at capturing sequential information, making them suitable for analyzing gene expression profiles that exhibit temporal dynamics within cells. Further, Autoencoders models can be used for dimensionality reduction and feature extraction from scRNA-seq data, potentially revealing hidden patterns or latent variables associated with DEGs [3].

Unsupervised Learning, such as Clustering algorithms (e.g. k-means, graph-based), can be employed to group cells based on their gene expression profiles, potentially revealing clusters enriched for DEGs associated with specific cell types or states, and dimensionality reduction techniques such as PCA and t-SNE can visualize high-dimensional scRNA-seq data in lower dimensions, aiding in the identification of DEGs associated with distinct cell populations. Clustering and dimensionality reduction techniques are already discussed in the previous Chapter. Some of the machine learning based approaches for DEA are listed in Table 4.1.

Table 4.1 List of machine learning based methods and tools for differential expression analysis

Methods/tools	Programming platform	Description	Code link	References
CloudPred	Python	A deep learning framework designed to predict cellular phenotypes	https://github.com/bryanhe/CloudPred	He et al. [8]
MRFscRNAseq	R	A Markov random field (MRF) model for network-based differential expression analysis to identify cell-type specific DEG	https://github.com/eddiehli/MRFscRNAseq	Li et al. [15]
scDEA	R	An ensemble learning-based method, combines 12 methods to produce more stable and accurate results	https://github.com/Zhangxf-ccnu/scDEA	Li et al. [14]
lvm-DE	Jupyter Notebook, R, Python	A generic Bayesian method for DEGs predictions from a fitted deep generative model	https://github.com/PierreBoyeau/lvm-DE-reproducibility	Boyeau et al. [2]
BOMA	R	A machine-learning framework that aligns developmental gene expression data between brains and organoids	https://github.com/daifengwanglab/BOMA	He et al. [9]

4.5 Case Studies

As discussed earlier, most of the methods, both statistical and machine learning, for DAE originally developed for RNA-seq have been repurposed for scRNA-seq. Development of machine learning based methods for DEGs identification, including cell type-specific DEGs, exclusively proposed for scRNA-seq are in its infancy. In this section, we present a few case studies of machine learning based DEA methods for scRNA-seq.

4.5.1 CloudPredict

Description: CloudPred, an interpretable machine learning algorithm proposed by He et al. [8], is designed to predict disease phenotypes from scRNA-seq data. Unlike traditional methods, CloudPred overcomes challenges such as varying cell counts and heterogeneous populations by employing an end-to-end differentiable learning approach and a biologically informed mixture of cell types model. Evaluation through both simulated and real dataset demonstrates CloudPred's superior performance in predicting clinical phenotypes and identifying relevant cell subpopulations, exemplified by its success in distinguishing lupus patients from controls with high accuracy.

Methodology: CloudPred is designed to automatically learn cell subpopulations whose variation across patients are indicative of phenotypes. It operates by modeling patient scRNA-seq data as mixtures of Gaussians, enabling estimation of cell population abundances per patient. These abundances serve as patient-specific features for phenotype prediction. Unlike traditional methods, CloudPred autonomously learns relevant cell populations without prior knowledge, training all model parameters end-to-end through stochastic gradient descent with a learning rate of 10^{-4} for 1000 epoch. Here, each patient is represented as a multiset $\left\{x_i \in \mathbb{R}^d\right\}_{i=1}^n$, where n represents the number of cells (points) from the patient, and d represents the dimension of the feature space (number of measured genes in scRNA-seq). The individual points are modelled as samples from a mixture of Gaussians, represented as m subpopulations characterized by shared parameters: mean (μ_j), covariance (ρ_j), and weight (w_i). To streamline parameter learning, covariance matrices are constrained to diagonal matrices. An Pytorch and Python based implementation of CloudPred can be found at https://github.com/bryanhe/CloudPred.

Results and Discussions: A real scRNA-seq dataset comprises of 142 lupus patients with 566,453 cells and controls were considered. The simulation results of CloudPred demonstrate efficacy in accurately predicting disease states (22 healthy, 120 lupus), race (80 European, 62 Asian), and monocyte composition estimated from Complete Blood Count (CBC). It achieves an AUROC of 0.98, identifying a distinct subset of CD4 T cells indicative of lupus. CloudPred has been shown to outperform other

Table 4.2 Performance evaluation of CloudPred and comparison with other methods on the lupus dataset as reported by He et al. [8]

Methods	AUROC		
	Lupus (disease)	Monocytes	Race
Mixture (Class)[a]	0.93	–	0.76
Mixture (Patient)[a]	0.77	–	0.64
Independent[b]	0.98	0.95	0.74
Deepset[c]	0.98	0.86	0.78
CloudPred	0.98	0.97	0.87

[a]An alternate method where two Gaussian mixture models are fitted separately for diseased patients and healthy controls. The likelihood of a new patient's scRNA-seq data under each model is calculated, and the phenotype with the higher likelihood is chosen for prediction. The model focusing on disease signatures is referred to as Mixture (class), while Mixture (patient) fits a Gaussian mixture to each patient's data individually
[b]This method treats cells from a patient as independent entities, predicting each cell's phenotype separately and then averaging these predictions
[c]Deep sets, a deep learning approach introduced by Zaheer et al. [21], is employed. It encompasses a broad range of permutation-invariant functions to make predictions on sets, with each set representing the scRNA-seq data of a patient

alternative methods in disease state prediction (Table 4.2). Predicting race proves more challenging, with CloudPred maintaining superior performance. Additionally, CloudPred successfully predicts monocyte composition, a feat not feasible with other methods relying on discrete classes. A detailed discussion on CloudPred can be found in He et al. [8].

4.5.2 MRFscRNAseq

Description: Li et al. [15] proposed a model, named MRF-scRNA-seq, that uses a Markov random field (MRF) to identify DEGs. This method takes into account gene networks and dependencies between cell types to identify genes that are differentially expressed in specific cell types. The method employs an Expectation–Maximization algorithm with a mean field-like approximation to estimate model parameters. Additionally, it utilizes a Gibbs sampler to identify DEGs.

Methodology: The simulation results of this method showed a better ability to identify cell-type specific DEGs compared to traditional methods while maintaining a low error rate. Li et al. [15] demonstrated the usefulness of this method by applying it to study idiopathic pulmonary fibrosis (IPF) utilizing a dataset of lung tissues from 32 IPF patients and 28 healthy controls. Out of 18,150 genes from 38 cell types, this study considers 2000 genes exhibiting high variations between cell types. The information

Table 4.3 DEGs identified under 3 different model settings with 3 types of test statistics on idiopathic pulmonary fibrosis datasets [15]

Test statistics	Different models settings	Number of DEGs identified in at least one cell-type	Aggregated number of cell-type specific DEGs
t-test	Main model	1472	3607
	Main with BioGrid	1605	4880
	Main with IntAct	1601	4875
MAST	MAST	1721	4826
	MAST with BioGrid	1870	8835
	MAST with IntAct	1867	8809
Wilcoxon	Wilcoxon	1562	3945
	Wilcoxon with BioGrid	1767	6526
	Wilcoxon with IntAct	1759	6474

related to gene networks was extracted from interactome databases BioGrid and IntAct. For these 2000 genes, the authors used two separate MRF models—one with gene networks from BioGrid, and the other with IntAct. The model parameters were estimated using the Expectation–Maximization (EM) algorithm over 200 iterations, followed by Gibbs sampler for 20,000 iterations, discarding the first 10,000 as burn-in, to obtain the posterior probabilities. This model has been implemented in the R package, and is available on GitHub https://github.com/eddiehli/MRFscRNAseq.

Results and Discussions: Among 2000 genes, the MRF model using the BioGrid gene network detected 1605 DEGs in at least one cell type, while using IntAct gene network the model identified 1601 DEGs. These results were compared with results from two-sample t-tests adjusted with Benjamini and Hochberg's procedure for false discovery rate (FDR), which identified 1472 DEGs. Other methods, like MAST, and Wilcoxon test have also been compared with MRF models (Table 4.3). One of the limitations of MRFscRNAseq method is that it lacks direct integration of gene expression directionality, making it difficult to discern up- or down-regulation of identified DEGs. A potential solution is to utilize original input test statistics' signs to infer expression changes and to enhance the model with weighted consideration, prioritizing transcription factors for their crucial role in gene regulation. Detailed discussion on the model can be found in Li et al. [15].

4.5.3 scDEA

Description: scDEA, proposed by Li et al. [14], is an ensemble learning method that combines the *p*-values from 12 individual DEA methods by using Lancaster's combined probability test to achieve consensus, accurate, and robust differential expression analysis. Comprehensive experiments demonstrate scDEA's superiority over some of the existing methods identifying biologically enriched DEGs and achieving concordance across datasets of varying sample sizes and experimental batches.

Methodology: scDEA is an ensemble method that combines 12 state-of-the-art DEA techniques for both bulk RNA-seq and scRNA-seq data. It employs a combined probability test to integrate *p*-values from these methods. The process involves normalizing raw read count matrices, running individual DEA methods, integrating *p*-values, and adjusting for multiple tests. This approach enhances accuracy and robustness in identifying DEGs. Once the *p*-values are obtained for each gene from 12 different DEA methods, it constructs a *p*-value matrix $P = (p_{jk}) \in R^{p \times 12}$, where p_{jk} denotes *p*-value for gene *j* computed by method *k*. Utilizing this *p*-value matrix, scDEA employs Lancaster's combined probability test, called Lancaster's test, to consolidate these *p*-values into a unified result. An R package https://github.com/Zhangxf-ccnu/scDEA and a Shiny application (https://github.com/Zhangxf-ccnu/scDEA-shiny) have been developed for user-friendly implementation and visualization.

Results and Discussions: In order to assess the performance of scDEA, six data sets were used. Its performance was compared with the 12 individual DEA methods on the basis of metrics like type I error control, gene ontology (GO) based enrichment analysis and sensitivity analysis of sample sizes and batch effects. Comparisons regarding type I error control demonstrate the superiority of scDEA over individual DEA methods. In GO analysis, scDEA also outperforms individual methods with more biological functions, and it identifies a larger number of DEGs. Additionally, investigations into impact of sample sizes and batch effects reveal scDEA's greater robustness compared to individual DEA methods. By leveraging a combined probability test method, scDEA achieves enhanced accuracy by amalgamating *p*-values from individual methods, thereby capitalizing on existing strengths rather than developing a new DEA method for scRNA-seq data.

4.5.4   lvm-DE

Description: Recent advancements in deep generative modeling have facilitated the development of methods for normalizing and analyzing gene expression data at scale, accommodating millions of cells and complex study designs. Most of the typical models represent gene expression in a cell as a nonlinear function of a low-dimensional latent summary of the cell's state and its batch identifier, enabling information sharing across cells and genes. However, rigorous and accurate DEA with these models have received little attention. Few methods utilize uncertainty in latent variables to approximate Bayes factors for DEGs detection, however, they have certain limitations including sensitivity in defining DEGs, challenges in interpreting Bayes factors in terms of established metrics like false discovery rate (FDR), and lacking theoretical justification for certain approaches, such as averaging Bayes factors from randomly sampled cell pairs. In order to overcome the mentioned issues, lvm-DE is developed by Boyeau et al. [2].

Methodology: lvm-DE [2] is a Bayesian approach leveraging deep generative models that outperform existing methods by estimating log fold changes and detecting DEGs while controlling FDR. The lvm-DE introduces a systematic model that enables the deployment of deep generative models to quantify and evaluate the significance of gene expression differences. This method accepts a latent variable model, a scRNA-seq dataset with covariates (e.g., batch identifiers), two cell groups for comparison, and a significance level α as input. It provides estimates of the log fold change (logFC) for each gene between the two groups and furnishes a list of significant DEGs controlling FDR at level α. These estimates are computed while adjusting for library size differences and variability due to observed technical confounders, such as batch effects. lvm-DE involves five steps:

   i. Fit a model to represent cell expression values using a low-dimensional latent variable.
  ii. Utilize annotations to construct posterior distributions of latent representations for each cell group.
 iii. Infer distributions of normalized gene expression for each cell group.
 iv. Estimate logFC and posterior probabilities for differential expression while accounting for batch effects.
  v. Identify significant genes based on desired FDR control.

The source code of lvm-DE method is available at https://github.com/PierreBoyeau/lvm-DE-reproducibility.

Results and Discussions: lvm-DE focuses on two primary tasks in DEA, (i) robust estimation of logFC in both noisy as well as sparse data, and (ii) evaluation of statistical significance for calling DEGs at a target FDR. Evaluation of the method was conducted with recently published methods including MAST, DESeq2, edgeR, and Limma-voom. Performance assessment for logFC estimation and prediction of DEGs sets is conducted using both synthetic and real datasets. Moreover, unlike most DEA pipelines, lvm-DE utilizes all available samples for between-type comparisons. Leveraging shared gene–gene correlations learned by latent variable models enhances gene expression modeling and increases detected DEGs. Thus, lvm-DE effectively characterizes gene expression differences in large-scale, batch-confounded datasets, offering calibrated DEGs based on posterior FDR expectations. While lvm-DE demonstrates broad applicability, two limitations should be noted. Firstly, slight modifications to existing latent variable models may be necessary for optimal performance. Empirically, it is reported that library size modeling to be crucial, especially in differential expression comparisons involving populations with varying RNA content. Secondly, lvm-DE may not be suitable for very small datasets with fewer than a few thousand cells, as fitting deep generative models may be challenging. In such cases, traditional methods that scale well and are less prone to overfitting are more appropriate.

4.5.5 BOMA

Description: Organoids serve as vital models for unraveling the intricacies of cellular and molecular mechanisms, notably in brain development. Yet, uncertainty looms over the preservation of gene expression between human organoids and brains, particularly in specific types of cells. Furthermore, a notable absence of efficient computational methods for comparative analyses of organoid and developing human brain data exists. To confront this challenge, He et al. [9] introduce a machine-learning framework for brain and organoid manifold alignment (BOMA). BOMA identifies conserved and specific developmental trajectories across human and non-human primate brains and organoids, shedding light on gene expression programs at cellular resolution.

Methodology: The BOMA pipeline facilitates the alignment of gene expression data from brain and organoid samples, with the aim of uncovering both conserved and specific developmental gene expression patterns. This enables a deeper exploration of developmental functional genomics at tissue and cell-type levels, particularly within organoids. To compare two developmental gene expression datasets (such as brains versus organoids), BOMA employs a two-step alignment process. Initially, it aligns brain and organoid samples globally pertaining to available timing information to establish an initial correspondence. In the second step, BOMA utilizes manifold learning for the refinement of alignment locally, co-embedding brain and organoid samples into a shared manifold space. The shapes of these manifolds reveal developmental trajectories, distinguishing between conserved (aligned) and brain/organoid-specific (unaligned) samples. Following sample alignment, BOMA utilizes Spectral clustering from Python's 'sklearn' package in order to cluster the aligned samples in common space based on their alignment scores. The number of clusters can be adjusted through various clustering parameters or by further sub-clustering existing coarse clusters. Additionally, BOMA identifies DEGs within clusters using Presto. This involves conducting both the Wilcoxon rank-sum test and AUROC analysis to compare cells within each cluster against all other cells in the dataset. The source code is available at https://github.com/daifengwanglab/BOMA.

Results and Discussions: Fig. 4.1 illustrates BOMA's process: it takes in developmental gene expression data from brain and organoid samples, aligns them globally to establish a coarse-grained correspondence matrix, then refines alignment through manifold alignment. It identifies shared manifolds and maps them to a common space, revealing developmental trajectories conserved across brains and organoids. BOMA clusters samples on these trajectories, identifying DEGs, and associated phenotypes for deeper insights. To showcase BOMA as a comparative analysis framework for brain and organoid development, He et al. [9] conducted various experiments. Initially, BOMA was applied to bulk-cell RNA-seq datasets, encompassing human brains, non-human primate brains, and human organoids. Additionally, the authors illustrated BOMA's effectiveness for the alignment of scRNA-seq datasets, integrating it from multiple independent studies. Furthermore, benchmarking of various contemporary alignment tools using these datasets has been done. The study demonstrates that BOMA's semi-supervised approach outperforms unsupervised methods for comparative analysis. The global alignment step enhances model interpretability and BOMA's ability to discover aligned developmental trajectories. BOMA significantly reduces confounders and shows better performance for integrative analysis of multiple studies. This method may easily integrate other alignment methods and can be incorporated into correspondences between sample pairs as prior knowledge. Experiments intentionally introducing mismatched brain regions and cell types illustrated the robustness of BOMA. The study also found similarities of gene expression between organoids and brains, indicating the sustainability of employing organoids for understanding the development of the human brain. The limitations include the assessment of BOMA on the limited number of samples and cultured periods. Future studies are encouraged to utilize longer-cultured organoids and a larger number of samples to enhance the comparative analysis.

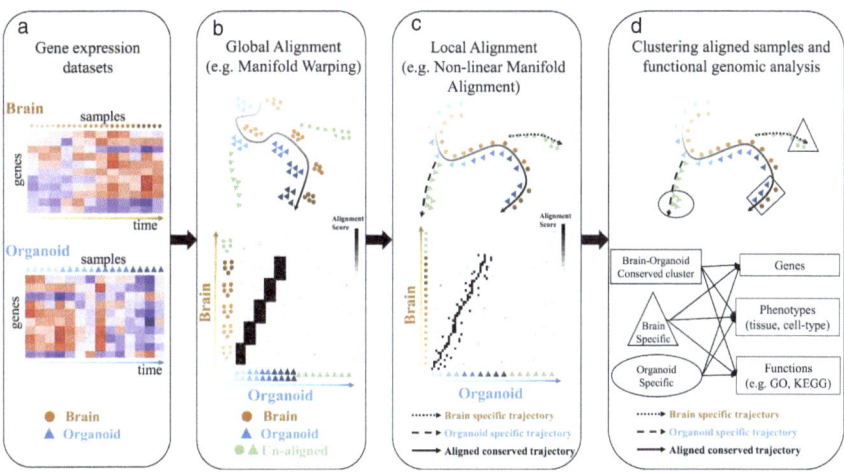

Fig. 4.1 The BOMA computational pipeline facilitates comparative analysis of developmental gene expression data between brains and organoids. **a** BOMA takes in multiple developmental gene expression datasets from brains and organoids, arranged according to prior timing information. **b** Step 1 involves global alignment for the purpose of inferring sample correspondences across different datasets at a coarse-grained level. **c** Step 2 refines alignment through local alignment, mapping samples onto a common manifold space. **d** Clustering and functional analysis of aligned samples reveal conserved (square) or specific (circle and triangle) clusters and developmental trajectories (black curves). Downstream analyses uncover differentially expressed genes, gene functions enrichment, and associated phenotypes, contributing to the understanding of developmental functional genomics. Adopted from He et al. [9]

4.6   Conclusion

The shift from traditional bulk RNA-seq to scRNA-seq has revolutionized gene expression analysis, allowing for a more detailed understanding of cellular heterogeneity and differential gene expression. Differential expression analysis (DEA) identifies genes that are actively expressed under different conditions, crucial for understanding diseases like cancer and developing therapeutics. Single-cell DEA offers enhanced sensitivity and resolution, uncovering rare cell types and states, and has widespread applications across biological disciplines, from cancer research to neuroscience. This chapter aimed to provide a comprehensive guide to the principles and methodologies of single-cell DEA, bridging theory with practical application to empower researchers in leveraging single-cell data effectively.

Various statistical methods have been developed for scRNA-seq DEA, addressing challenges like sparsity and dropout events. DESeq2 and edgeR, originally for bulk RNA-seq, have been adapted for scRNA-seq, employing negative binomial models. MAST, SCDE, and BASiCS offer specialized approaches for scRNA-seq data, considering single-cell variability and dropout events. IDEAS uniquely analyzes scRNA-seq data across multiple individuals, estimating individual-specific expression distributions to identify DEGs. Selection of the method depends on data characteristics and experimental design, with detailed discussions available elsewhere.

Machine learning approaches, including CNNs, RNNs, and Autoencoders, are pivotal for scRNA-seq DEA, enabling the identification of DEGs at the single-cell level. Unsupervised learning techniques like clustering algorithms aid in grouping cells based on expression profiles, while dimensionality reduction methods like PCA and t-SNE visualize scRNA-seq data, aiding DEGs identification. Notable machine learning-based methods for DEA discussed in this Chapter include CloudPred for predicting disease phenotypes, MRFscRNAseq for identifying cell-type-specific DEGs, scDEA for ensemble learning-based DEA, lvm-DE for Bayesian modeling of DEGs, and BOMA for aligning brain and organoid developmental trajectories. These methods offer robust frameworks for analyzing scRNA-seq data, enhancing our understanding of gene expression dynamics in complex biological systems.

References

1. N. Bano, I. Sajid, S.A.A. Faizi, A. Mutshembele, D. Barh, K. Raza, Computational intelligence methods for biomarkers discovery in autoimmune diseases: case studies, in *Studies in Computational Intelligence*, vol. 1133 (Springer, 2024)
2. P. Boyeau, J. Regier, A. Gayoso, M.I. Jordan, R. Lopez, N. Yosef, An empirical Bayes method for differential expression analysis of single cells with deep generative models. Proc. Natl. Acad. Sci. U.S.A. **120**(21), e2209124120 (2023). https://doi.org/10.1073/pnas.2209124120
3. M. Brendel, C. Su, Z. Bai, H. Zhang, O. Elemento, F. Wang, Application of deep learning on single-cell RNA sequencing data analysis: a review. Genom. Proteomics Bioinform. **20**(5), 814–835 (2022). https://doi.org/10.1016/j.gpb.2022.11.011

4. Y. Chen, D. McCarthy, M. Robinson, G.K. Smyth, edgeR: differential expression analysis of digital gene expression data user's guide, in *Bioconductor User's Guide* (2014)

5. S. Das, S.N. Rai, SwarnSeq: an improved statistical approach for differential expression analysis of single-cell RNA-seq data. Genomics 113(3), 1308–1324 (2021)

6. S. Das, A. Rai, S.N. Rai, Differential expression analysis of single-cell RNA-seq data: current statistical approaches and outstanding challenges. Entropy (Basel, Switz.) 24(7), 995 (2022). https://doi.org/10.3390/e24070995

7. G. Finak, A. McDavid, M. Yajima, J. Deng, V. Gersuk, A.K. Shalek, R. Gottardo et al., MAST: a flexible statistical framework for assessing transcriptional changes and characterizing heterogeneity in single-cell RNA sequencing data. Genome Biol. 16, 1–13 (2015)

8. B. He, M. Thomson, M. Subramaniam, R. Perez, C.J. Ye, J. Zou, Cloudpred: predicting patient phenotypes from single-cell RNA-seq, in *Pacific Symposium on Biocomputing 2022* (2021), pp. 337–348

9. C. He, N.C. Kalafut, S.O. Sandoval, R. Risgaard, C.L. Sirois, C. Yang, D. Wang, et al. BOMA, a machine-learning framework for comparative gene expression analysis across brains and organoids. Cell Rep. Methods 3(2) (2023)

10. A. Jabeen, N. Ahmad, K. Raza, Machine learning-based state-of-the-art methods for the classification of RNA-seq data, in: *Classification in BioApps*, ed. by N. Dey, A. Ashour, S. Borra. Lecture Notes in Computational Vision and Biomechanics, vol. 26 (Springer, 2018), pp. 133–172

11. A. Jabeen, N. Ahmad, K. Raza, Global gene expression and docking profiling of COVID-19 infection. Front. Genet. 13, 870836 (2022)

12. F.N. Khan, M. Asim, M.I. Qureshi, Artificial intelligence in the diagnosis and treatment of rheumatoid arthritis: current status and future prospects, in *Artificial Intelligence and Autoimmune Diseases*, ed. by K. Raza, S. Singh. Studies in Computational Intelligence, vol 1133 (Springer, 2024)

13. P.V. Kharchenko, L. Silberstein, D.T. Scadden, Bayesian approach to single-cell differential expression analysis. Nat. Methods 11(7), 740–742 (2014)

14. H.S. Li, L. Ou-Yang, Y. Zhu, H. Yan, X.F. Zhang, ScDEA: differential expression analysis in single-cell RNA-sequencing data via ensemble learning. Brief. Bioinform. 23(1), bbab402 (2022). https://doi.org/10.1093/bib/bbab402

15. H. Li, B. Zhu, Z. Xu, T. Adams, N. Kaminski, H. Zhao, A Markov random field model for network-based differential expression analysis of single-cell RNA-seq data. BMC Bioinform. 22, 1–16 (2021)

16. M. Love, S. Anders, W. Huber, Differential analysis of count data–the DESeq2 package. Genome Biol. 15(550), 10–1186 (2014)

17. A. Sahu, S. Qazi, K. Raza, A. Singh, S. Verma, Machine learning-based approach for early diagnosis of breast cancer using biomarkers and gene expression profiles, in *Computational Intelligence in Oncology, Studies in Computational Intelligence (SCI)*, vol. 1016 (Springer, 2022), pp. 285–306

18. S.M. Swain, M. Shastry, E. Hamilton, Targeting HER2-positive breast cancer: advances and future directions. Nat. Rev. Drug Discov. 22(2), 101–126 (2023). https://doi.org/10.1038/s41573-022-00579-0

19. C.A. Vallejos, J.C. Marioni, S. Richardson, BASiCS: Bayesian analysis of single-cell sequencing data. PLoS Comput. Biol. 11(6), e1004333 (2015)

20. L. Yang, A.M. Bhat, S. Qazi, K. Raza, DLC1 as Druggable target for specific subsets of gastric cancer: an RNA-seq-based study. Medicina 59(3), 514 (2023)

21. M. Zaheer, S. Kottur, S. Ravanbakhsh, B. Poczos, R.R. Salakhutdinov, A.J. Smola, Deep sets, in *Advances in Neural Information Processing Systems*, vol. 30 (2017)

22. M. Zhang, S. Liu, Z. Miao, F. Han, R. Gottardo, W. Sun, IDEAS: individual level differential expression analysis for single-cell RNA-seq data. Genome Biol. 23(1), 33 (2022)

Chapter 5
Trajectory Inference and Cell Fate Prediction

5.1 Introduction

Living cells do not remain static; they undergo changes in their gene expression profiles as they move through different stages of their lifecycle or in response to environmental cues. A trajectory captures these changes, illustrating the sequence of transitions from one cell state to another. In the context of scRNA-seq data, a trajectory refers to the inferred path that living cells follow as they progress through different states over time during a dynamic biological process, such as development, differentiation, or response to stimuli [7, 18]. Cell fate prediction involves forecasting the future states or identities of cells based on their current gene expression profiles. It's like predicting a stem cell's future career path within the body. It uses computational models and algorithms to predict how individual cells will differentiate or respond to various stimuli over time. It leverages the trajectories inferred from scRNA-seq data to predict how individual cells will differentiate or respond to various stimuli over time [5].

Trajectory inference and cell fate prediction have a wide range of applications in medical sciences, offering valuable insights into cellular development, disease mechanisms, regenerative medicine, personalized treatments, drug development, biomarker discovery, basic research, and immunology. These tools enable researchers and clinicians to better understand and manipulate cellular processes, leading to advances in the diagnosis, treatment, and prevention of diseases. In addition to scRNA-seq data preprocessing and feature extraction, machine learning also plays a crucial role in trajectory inference, and cell fate prediction.

There are both separate and common software tools for trajectory inference and cell fate prediction. While trajectory inference often serves as a basis for cell fate prediction, some tools are specifically designed for one task or the other, and others integrate both functionalities. Commonly used methods for trajectory inference include Slingshot [17], Monocle [3], PAGA [21], VITAE [4], and scTEP [22], construct trajectory graphs to map cell progression through different states. Also,

K. Raza, *Machine Learning in Single-Cell RNA-seq Data Analysis*,
SpringerBriefs in Computational Intelligence,
https://doi.org/10.1007/978-981-97-6703-8_5

probabilistic models such as Gaussian processes and hidden Markov models (HMMs) capture uncertainty in cell state transitions, providing a probabilistic framework for trajectory inference. Commonly used tools for cell fate and state prediction, and mapping include SCENIC [1], CellRank [8], and CellOracle [6]. Integrated tools used for both trajectory and cell fate prediction are Scanpy [20], VIA [16], and scTour [10].

Similarly, supervised learning with models like random forests, SVMs, and ANNs predict future cell states from current gene expression profiles. Further, deep learning with Convolutional Neural Networks (CNNs), Recurrent Neural Networks (RNNs), and autoencoders capture complex temporal and spatial patterns in gene expression data to enhance cell fate prediction accuracy [15]. Most of the existing trajectory inference methods often struggle with handling multiple biological conditions. To address this, the method "condiments" [11] has been introduced for inferring and interpreting cell trajectories across different conditions, such as comparing wild-type and knock-out stem cell populations. Condiments integrate datasets from multiple conditions into a single trajectory and identify large-scale changes in differentiation progression or fate selection. It also detects subtler differences in gene expression behaviours along the trajectory, providing comprehensive insights into condition-specific developmental processes.

The popularly used methods for trajectory inference and cell fate prediction are listed in Table 5.1, and some of them have been discussed in the following sections.

5.2   Trajectory Inference Methods

Trajectory inference (TI) is a computational method used to understand the dynamic processes of cellular development and differentiation using scRNA-seq data. It aims to reconstruct the progression of cell states over time, often represented as a trajectory or lineage tree, which shows how cells evolve from one state to another. The key steps in trajectory inference are data preprocessing (normalizing, filtering, etc.), dimensionality reduction, trajectory construction to infer the trajectory or pseudo-time, which is a latent variable representing the progression of cells through developmental stages, and branch identification for detecting points where cells diverge into different lineages or fates [12].

Trajectory inference calculates the pattern of gene expression changes in cells within a dataset and arranges them in a pseudo-chronological order along a developmental pathway (pseudotime) based on similarities in their gene expression profiles [7]. Trajectory inference methods infer a graph-like structure of the dynamic process, mapping cells to this structure to analyze their properties over pseudotime. This approach reveals how cells transition between states and make fate decisions [18]. Pseudotime is an abstract measure that represents the progression of cells along the trajectory. It orders cells not by actual chronological time but by their progression

Table 5.1 A list of computational tools for trajectory inference and cell fate predictions

Tools	Methods	Descriptions	Programming platform	Link	References
SCENIC	Integrates 3 tools GENIE3, RcisTarget, and AUCell	Reconstruction of GRNs and cell states identification	Python, R	http://scenic.aertslab.org	Aibar et al. [1]
Scanpy	Utilizes existing Python packages and libraries	A toolkit for preprocessing, clustering, pseudotime and trajectory inference, DEA, and simulation of GRNs	Python	https://github.com/theislab/Scanpy	Wolf et al. [20]
Slingshot	Cluster-based minimum spanning tre	Infers cell lineages and pseudotimes	R	https://github.com/kstreet13/slingshot	Street et al. [17]
Monocle	Reverse-graph embedding	A toolkit for clustering, cell classification, trajectory inference, and pseudotime estimation	R	https://github.com/cole-trapnell-lab/monocle3	Cao et al. [3]
PAGA	Integrates graph abstraction and partitioning methods	Infers developmental trajectories and cell differentiation pathways	Python	https://github.com/theislab/paga	Wolf et al. [21]
Palantir	Diffusion maps, and Markov chain	Model trajectories of differentiating cells	Python	https://github.com/dpeerlab/Palantir/	Setty et al. [13]
scVelo	Likelihood-based dynamical model	Infers gene-specific transcription, splicing, and degradation rates; finds each cell's differentiation stage and driver genes	Python	https://scvelo.org/	Bergen et al. [2]
VITAE	Combines latent hierarchical mixture model with variational autoencoders	Joint trajectory inference	Python	https://github.com/jaydu1/VITAE	Du et al. [4]

(continued)

Table 5.1 (continued)

Tools	Methods	Descriptions	Programming platform	Link	References
VIA	Lazy-teleporting random walks with Markov chain Monte Carlo refinement	Generalized and scalable method to compute pseudotime, and reconstruct cell lineages	Python	https://github.com/ShobiStassen/VIA	Stassen et al. [16]
CellRank	Combines Markov chain modeling with RNA velocity	Detects initial, intermediate, and terminal cell populations, predict fate potentials, and visualize gene expression trends along lineages	Python	https://github.com/theislab/cellrank	Lange et al. [8] and Weiler et al. [19]
CellOracle	Combines GRN reconstruction and machine learning based methods	Gene perturbation analyses, simulating changes in cell identity	Python	https://github.com/morris-lab/CellOracle	Kamimoto et al. [6]
scTour	Combines variational autoencoders and neural ordinary differential equation	Simultaneous inference of developmental pseudotime, transcriptomic vector field, and latent space of cells	Python	https://github.com/LiQian-XC/sctour	Li [10]
scTEP	Autoencoder and Minimum Spanning Tree	Trajectory inference method using Ensemble Pseudotime inference	R	https://cran.r-project.org/package=scTEP	Zhang et al. [22]

through the biological process being studied. Cells at the beginning of the trajectory have low pseudotime values, while those further along the path have higher pseudotime values.

A trajectory can have branches, indicating points where cells diverge into different lineages or fates. For example, a stem cell might differentiate into several types of specialized cells, each following a distinct branch of the trajectory. Branching points help identify key decision-making events in cellular differentiation. A multitude of

TI methods exist, with varying characteristics such as fixed versus inferred topology and supported graph topologies. While early methods focused on ordering cells along a fixed trajectory, recent ones infer the trajectory topology. Quantitative assessment of TI methods' performance, scalability, robustness, and usability is crucial. A comprehensive comparison of 45 TI methods over diverse datasets has been done by Saelens et al. [12]. The primary difference among TI methods lies in whether they fix the topology and, if not, the types of topology they can detect. Common topologies include linear, cyclical, and bifurcating, with more complex structures like connected and disconnected graphs also considered. Many methods prioritize inferring linear trajectories or simpler tree-like topologies, while only a few explore cyclic or disconnected structures.

5.2.1   Monocle 3

Monocle 3 is a computational tool used for trajectory inference and analysis of scRNA-seq data. It is developed by the Trapnell Lab at the University of Washington. Monocle 3 builds upon the previous versions (Monocle 1 and Monocle 2) and introduces several new features and improvements. Some of the key features of Monocle 3 are (i) graph-based trajectory inference, (ii) topology inference, (iii) scalability, (iv) integration with single-cell experiment frameworks, and (v) visualization tools.

Monocle 3 primarily utilizes a graph-based approach to learn trajectories from scRNA-seq data. Specifically, it employs the reverse graph embedding algorithm to infer trajectories of gene expression changes in each cell, placing them along the trajectory accordingly. Monocle's differential analysis toolkit allows for the identification of genes regulated over pseudotime, and in cases of multiple outcomes, it reconstructs branched trajectories corresponding to cellular decisions. This algorithm constructs a graph representing the relationships between cells based on their gene expression profiles and then embeds the cells into a low-dimensional space, such that their positions in the space reflect their progression along the trajectory. This embedding process allows Monocle 3 to order the cells along the trajectory and identify branching points corresponding to different cell fate decisions. Additionally, Monocle 3 incorporates machine learning techniques and statistical methods to improve the accuracy and robustness of trajectory inference.

5.2.2   PAGA

PAGA (Partition-based Graph Abstraction) is a computational framework for inferring developmental trajectories and cell differentiation pathways, developed by Wolf et al. [21]. PAGA constructs and analyzes graphs representing relationships between cells based on their gene expression profiles. The working steps of PAGA are described as follows:

Data Preprocessing: This step includes basic data filtering, total count normalization, log transformation, identification of highly variable genes, potential regression of confounding factors, and scaling to z-scores. The processed data is then subjected to PCA to reduce dimensionality within the reduced space of principal components.

Graph Construction: A symmetrized kNN-like graph is constructed using the compressed and denoised representation of the data, often employing UMAP's approximate nearest neighbor search. The graph typically utilizes Euclidean distance and may be weighted using adaptive Gaussian kernels or the exponential kernel within UMAP, with the latter being the choice for the presented results.

Graph Partitioning and Abstraction: All relevant partitioning of the kNN-like graph is considered. PAGA utilizes the Louvain algorithm to generate a PAGA graph for each partitioning, utilizing the "PAGA connectivity measure" to quantify connectivity between clusters, akin to modularity. In the directed case, involving a velocity graph derived from RNA velocity, the ratio of arrows between partitions is considered to assess transition tendencies.

Trajectory Inference: PAGA infers trajectories and developmental pathways by analyzing the abstracted graph. By identifying connected components and paths within the graph, it reveals potential differentiation trajectories and transitions between cell states.

5.2.3 scTEP

scTEP (Single-cell data Trajectory inference method using Ensemble Pseudotime inference) is a trajectory inference framework proposed by Zhang et al. [22]. It uses results of multiple clustering for inferring robust pseudotime and fine-tunes the trajectory accordingly. Evaluation on 41 real scRNA-seq datasets with ground truth trajectories shows that scTEP outperforms many existing methods.

The scTEP comprises four main components:

Pathway gene set intersection: Uses pathway information to generate latent representations for all pathways.

scDHA clustering and dimension reduction: Incorporates a method involving non-negative kernel autoencoder and variational autoencoder (VAE), to achieve superior performance in latent representation and clustering tasks.

Pseudotime inference from multiple clustering results: It derives pseudotime from the outcomes of multiple clusters to produce more resilient pseudotime estimates.

Pseudotime fine-tuned trajectory inference: Utilizes inferred pseudotime to fine-tune the constructed graph by sorting vertices according to their average pseudotime.

5.3 Cell Fate Prediction Methods

Cell fate prediction is the process of inferring the future developmental trajectory or ultimate fate of individual cells based on their gene expression profiles. This involves identifying the cellular state or differentiation pathway that a cell is likely to follow. It's important because it provides insights into how cells develop, differentiate, and specialize, which is fundamental for understanding various biological processes such as embryonic development, tissue regeneration, and disease progression. By predicting cell fate, researchers can uncover key regulatory mechanisms underlying cellular decision-making and potentially identify therapeutic targets for treating diseases. Cell fate determination is a highly conserved process across evolution, enabling cells to make crucial decisions about their roles within multicellular organisms. It is fundamental for orchestrating normal development, maintaining internal balance (homeostasis), and supporting tissue regeneration in adults. This process is remarkably robust, withstanding various disturbances due to its critical importance [9].

5.3.1 SCENIC

SCENIC (Single-Cell rEgulatory Network Inference and Clustering) [1] addresses the challenge of interpreting scRNA-seq data by simultaneously reconstructing gene regulatory networks (GRNs) and identifying cell states. It maps GRNs and identifies stable cellular states by evaluating the activity of these GRNs in each cell. The computational workflow of SCENIC involves three main steps: (i) identifying co-expressed gene sets using GENIE3, refining these modules with cis-regulatory motif analyses using RcisTarget, and scoring the activity of regulons in each cell using AUCell, which are available in R/Bioconductor package. Applied to diverse single-cell datasets from tumors and the brain, SCENIC reveals how the genomic regulatory code guides the identification of transcription factors (TFs) and cell states. This approach is considered robust against dropouts, as it scores the activity of entire regulons rather than individual genes or TFs.

5.3.2 CellRank

CellRank [8, 19] is a computational tool designed for single-cell fate mapping in diverse biological scenarios, such as regeneration, reprogramming, and disease, where the direction of the process is unknown. It combines trajectory inference with RNA velocity to account for the gradual and stochastic nature of cellular fate decisions and uncertainties in velocity vectors. CellRank can automatically detect initial, intermediate, and terminal cell populations, predict fate potentials, and visualize gene expression trends along lineages. Applied to pancreas development data, it effectively identifies cell states and fate potentials. In cellular reprogramming data, it accurately predicts reprogramming outcomes.

5.3.3   CellOracle

CellOracle [6] is a machine-learning-based tool that uses gene regulatory networks (GRNs) inferred from single-cell multi-omics data to simulate the effects of transcription factor (TF) perturbations on cell identity, using only unperturbed wild-type data. Applied to mouse and human haematopoiesis and zebrafish embryogenesis, CellOracle accurately models known phenotypic changes resulting from TF perturbations. In zebrafish development, it predicts and experimentally validates a new phenotype caused by the loss of the noto TF and identifies LHX1A as an axial mesoderm regulator. These findings demonstrate CellOracle's utility in inferring and interpreting cell-type-specific GRN configurations, providing mechanistic insights into cell identity regulation.

5.4   Deep Learning Framework for Inferencing Cellular Dynamics

A recently developed deep learning architecture by Li [10], named scTour, addresses the need for a batch-insensitive tool capable of inferring and predicting developmental dynamics. ScTour provides strong inference capabilities and precise predictions of cellular dynamics, effectively mitigating batch effects. In inference tasks, it concurrently estimates developmental pseudotime, defines the vector field, and maps the transcriptomic latent space within a unified framework. For prediction purposes, it reliably reconstructs the dynamics of unseen cellular states or entirely new independent datasets. Its functionalities are demonstrated across various biological processes using 19 datasets. It does not require a starting cell for pseudotime estimation. The low-dimensional latent space combines intrinsic transcriptome and extrinsic time information, enriching cell trajectory reconstruction. scTour's enables unbiased integration of different datasets. Additionally, it can predict transcriptomic properties and dynamics of unseen cellular states and new datasets.

The scTour architecture combines VAE and neural ordinary differential equation (ODE). It introduces a neural network to assign developmental time to each cell. This method leverages two encoder networks to capture hidden information (z-space) from a gene expression matrix. One network focuses on gene patterns, while the other estimates a developmental timeline (pseudotime) for each cell. These captured features, along with initial cell states and pseudotime points, are fed into a special neural network (neural ODE) to create a series of cell representations across the developmental trajectory. Finally, a decoder network uses these representations to reconstruct the original gene expression data. This unsupervised approach allows researchers to infer pseudotime, the underlying developmental changes (vector field), and hidden cellular states (Fig. 5.1). Additionally, the model can predict cellular dynamics of unobserved transcriptomes or time intervals. Additionally, scTour utilizes standard mini-batch training for efficiency and scalability. It offers two main functionalities: inference and prediction. For inference, it estimates cell pseudotime and transcriptomic vector field without relying on RNA velocity-based methods. For prediction, it can predict pseudotime, transcriptomic vector field, and latent space of unobserved cellular states or new datasets (Fig. 5.1). These features enhance scTour's utility in deciphering cellular dynamics in a batch-insensitive manner.

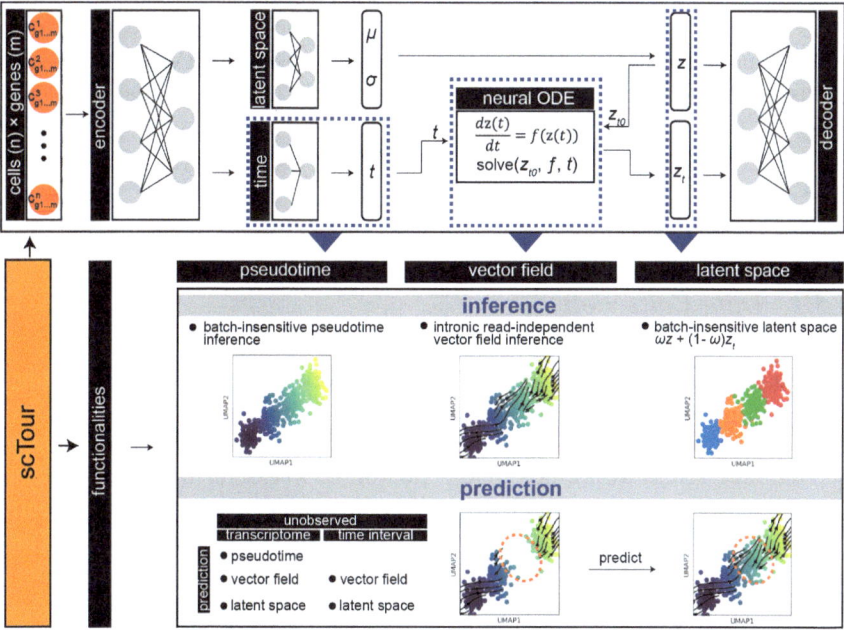

Fig. 5.1 The scTour framework utilizes encoder networks to generate distribution parameters for the approximate posterior (latent space, z) and assign time points to each cell (time, t) based on input gene expression matrix. The decoder network uses these latent representations to reconstruct input, inferring developmental pseudotime, transcriptomic vector field, and cell latent representations. Additionally, it predicts cellular dynamics for unobserved transcriptomes or time intervals. *Image credit* Li [10]. Reproduced under CC 4.0

Table 5.2 Unique cell identifiers for each patient across time courses

Patient code (GEO dataset)	Patient ID	Progression duration (early to late tumor)	Time points (TP)/cell barcodes			
			TP3	TP4	TP5	TP6
			Early tumor	Transition tumor		Late tumor
Patient 2	P1	4–12 weeks	1314	1294	549	580
Patient 5	P2	4–9 weeks	647	5893	1669	2213
Patient 6	P3	4–8 weeks	3729	3635	4603	2469
Patient 7	P4	5–11 weeks	3381	5468	4232	3865

5.5   Case Studies: Trajectory Inference via PAGA

In this section, we study on analyzing scRNA-seq data from aggressive neuroen-docrine small cell prostate cancer trans-differentiation using the PAGA software tool. In this analysis, we used Jupyter Notebook version 6.5.4 in an Anaconda environment for exploratory data analysis, data cleaning, visualization, and trajectory inference. We employed several specialized Python libraries, including Scanpy, ScVelo, iGraph, and Cellxgene.

5.5.1   Data Collection

The data was collected from NCBI-GEO with Accession No. GSE240056 (https://www.ncbi.nlm.nih.gov/geo/query/acc.cgi?acc=GSE240056). The study focused on analyzing trans-differentiation of tumor from adenocarcinoma to small cell neuroen-docrine state across a specified time period of 12 weeks for four patients. Gene expression (read counts) matrix has cell barcodes as columns that uniquely identify cells, and the rows represent read counts. Each cell barcode contains patient and time point information. Cancer progression across patients, time points (TPs), labeling of each TPs (early, transition, late), and cell barcodes are presented in Table 5.2.

5.5.2   Data Preprocessing

Normalization was performed to remove technical biases from the data. Batch effects, introduced by varying experimental conditions, were removed using Seurat (for R) and scScope (for Python), ensuring that only biologically relevant differences remained in the data.

5.5.3 Dimensionality Reduction, Clustering, and Visualization

As scRNA-seq data is large, it is required to reduce its dimension before analysis. Dimensionality reduction can be linear as in PCA or non-linear as seen in UMAP. Linear dimensionality reduction retains variations in the data while non-linear converts high dimensional dataset to low dimensional 2D space. Evaluation metrics are small intra-cluster distances and large inter-cluster distances. This signifies each cluster has highly similar components and different clusters have significant distances between them for clear interpretation. Feature selection is applied to find a set of highly informative genes from the entire dataset that can be used for further analysis.

Plotting UMAP for cell barcode labels provides an understanding of cancer progression in each patient and how it differs based on individual patient metabolism. Here, we analyze cancer growth for the time period of 11–12 weeks in four patients (P1, P2, P3, and P4) as depicted in Fig. 5.2. P1 has 3737 cells, therefore the color scale reaches a little above 3500, similarly, P2 has 11,421 so the scale is above 10,000, and the same follows for P3 and P4. From Fig. 5.2, we inferred that P1 and P2 have distinct gene expressions for each stage (TP3, TP4, TP5, TP6). The only difference is that P1 has a higher number of cells in the early stage (TP3) than in the last stage of cancer (TP6). Whereas, in P2 genes are highly expressed in transition state (TP4) cells than in any other state. But even here, there is clear discrimination between each stage and minimal overlap. This distribution is drastically different in P3 and P4. This signifies the same genes are expressed in all patients but distinctly in each tumor stage. Understanding the distribution of cell barcodes aids us in understanding an overview of how cancers spread across various patients and provides a lead that we should focus on detecting specific genes expressed in the clusters. This will in turn help us detect how the expression of specific genes impacts cancer growth over time in different individuals.

UMAP plot with 3 clusters signifying each stage of cancer growth is depicted in Fig. 5.3. From the Leiden clusters, we group together similar clusters into three final clusters to visualize tumor growth in all four samples. We observe that although P1 and P2 have clear boundaries between each stage, P3 and P4 have significant overlap between stages signifying genes that are co-expressed in multiple cell stages. Moreover, we understand that P3 shows significantly different growth compared to all other samples with minimum cells in the early stage and common genes contributing to transition and late stage of the tumor. P4 shows a clear division between the early stage of cancer and all other stages. We may conclude that there are very few genes expressed in the early stage of cancer and a maximum number of genes expressed in the transition stage, therefore for disease prevention and diagnosis, genes expressed highly in this stage can be further studied.

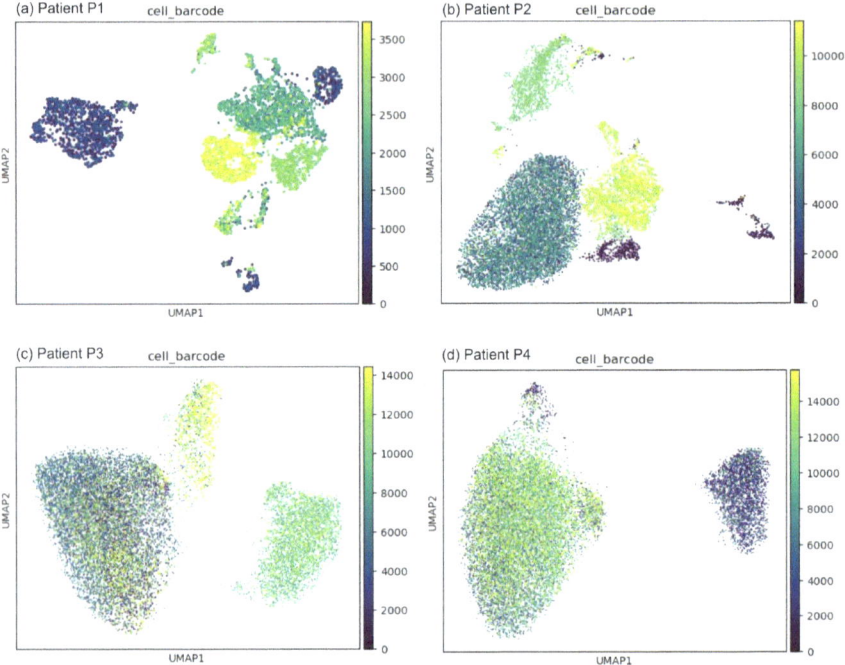

Fig. 5.2 UMAP analysis of each patient across cell barcode distribution, providing an estimate of cancer progression. Cell barcodes are labeled starting from 0 till the end and color scale on the right gives a description of the cell barcodes. **a** In patient P1, we can clearly see distinct clusters for each stage of cancer progression with dark-colored clusters close to 0 cell barcode index signifying early tumor and lightest-colored cluster showing late tumor cluster. **b** In P2, we get 4 distinctly colored clusters signifying each stage of cancer and here early-stage cancer is minimal while the transition stage is maximum. **c** In P3, only 3 clusters are observed where the early stage and transition stage cells are mixed in a single cluster while the last stage is separate. **d** In P4, only 2 clusters are observed which signifies genes for early-stage cancer are distinct while cells for transition and late-stage cancer have similar gene expression

5.5.4 Trajectory Inference via PAGA

Inferring cell–cell transitions and cellular heterogeneity was previously achieved through clustering data into discrete groups of similar expression levels or generating tree-based topology through time point data. However, the trajectories plotted through these methods were either disconnected or not capable of depicting enough branches in the tree. These limitations arose due to the inability of the algorithm to predict the range of changes showcased by single-cell data or the lack of sufficient biological data to map routes accurately. PAGA is a graph-based technique that covers both connected and disconnected trajectories. It utilizes kNN graphs to find connections between clusters and find a path that covers all cell clusters efficiently integrating branches and disconnected trajectories. It assigns a node to each cell,

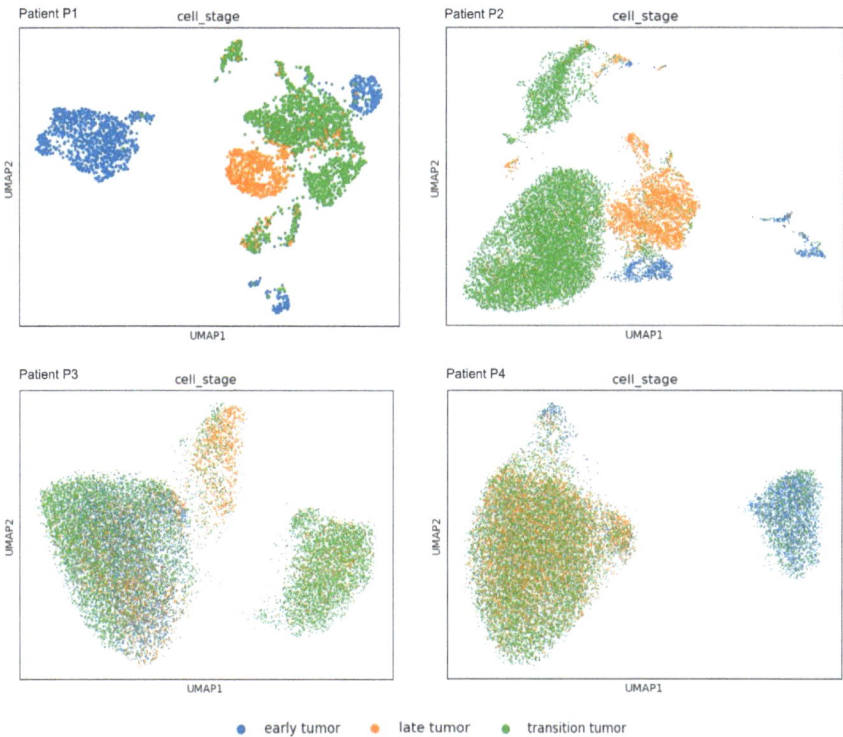

Fig. 5.3 UMAP plot showing stages of cancer growth (early, late, and transition tumor) for patients P1, P2, P3, and P4

and the edges represent connections with the edge weight signifying the strength of the connection. PAGA allows faster, more accurate, and inclusive cell lineages to be predicted.

We performed the trajectory analysis for all four patients using PAGA, utilizing Leiden clusters and 3 stages of cancer clusters, shown in Fig. 5.4. Based on the trajectory we infer connectivity among nodes and strengths of edges connecting the nodes. Here, we analyze the path toward the endpoint of tumor growth among all patients and specific nodes that signify clusters that have high connectivity among them. It helps detect which clusters have strong associations between them and how they differ in all samples. Further, this information can be used as labels to train machine learning models to predict cell fate.

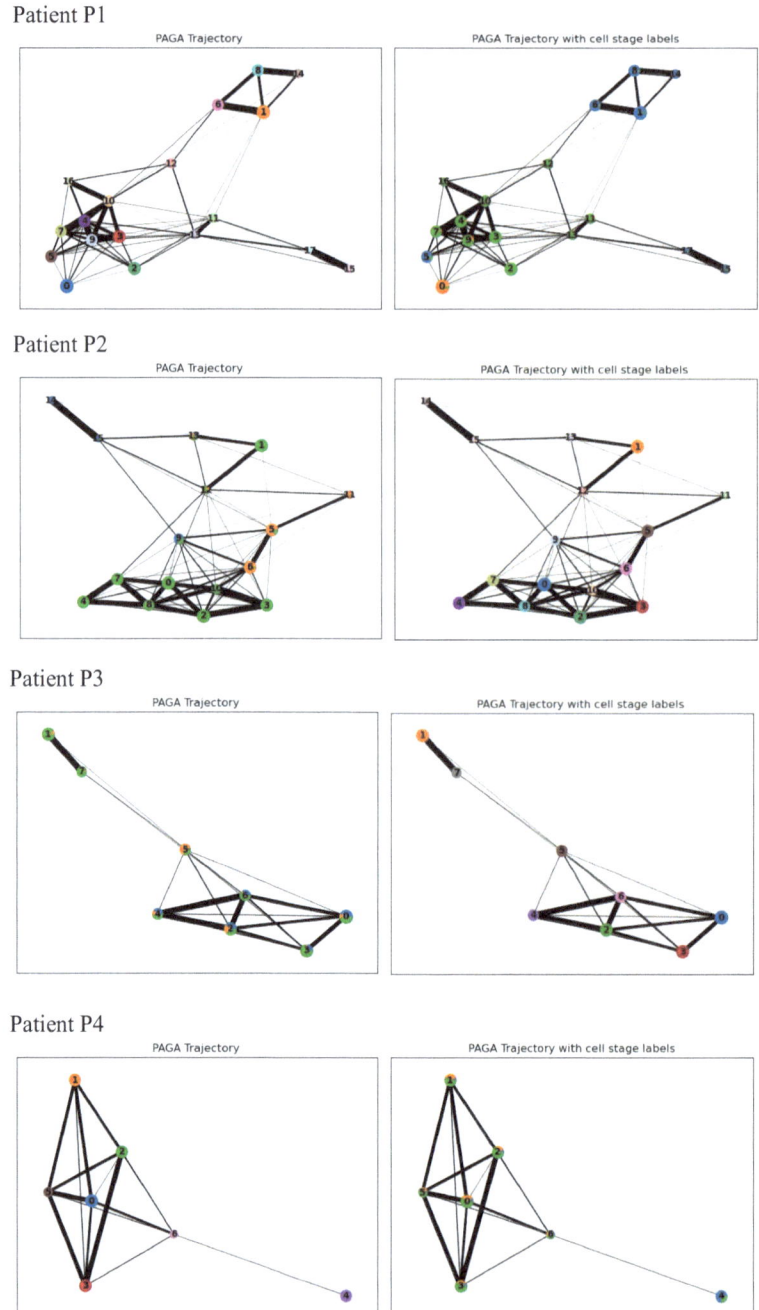

Fig. 5.4 Trajectory analysis for all four patients (P1, P2, P3, and P4)

5.6 Challenges and Future Directions

Trajectory inference and cell fate prediction using scRNA-seq are powerful techniques for understanding cellular dynamics and differentiation, however, these methods face several challenges that need to be addressed by researchers.

Technical and biological noise

Technical variability (dropout events, batch effects), and biological variability (cell heterogeneity, dynamic processes) are primary technical problems in scRNa-seq. In dropout events, certain mRNA molecules are not captured, leading to zero counts for genes that are actually expressed. Batch effects arise from differences in sample handling, library preparation, sequencing depth, and other factors, that can obscure true biological differences. Biological variability within a population of cells can be high, reflecting the diverse states and environments cells can be in. This intrinsic heterogeneity can complicate the identification of meaningful trajectories. Further, cellular processes are dynamic and can occur at different rates. Capturing cells at a single time point may not provide enough information to accurately infer trajectories.

Data sparsity and high dimensionality

scRNA-seq data are inherently sparse, that is, many genes have zero read counts in many cells. This sparsity can make it difficult to identify true signals from noise. Also, high dimensionality (thousands of genes) poses computational challenges and requires dimensionality reduction techniques, which may lose important information.

Choice of the model

While performing scRNA-seq analysis, including trajectory inference and cell fate prediction, we have linear versus non-linear models, and deterministic versus stochastic models. Several traditional methods assume linearity in gene expression changes, which might not capture the complexity of biological processes. However, non-linear methods are more flexible, but they are computationally intensive and harder to interpret. Deterministic models may not account for the inherent stochastic nature of gene expression, while stochastic models can address this but at the cost of increased complexity.

Algorithmic scalability and big data challenges

Trajectory inference and other scRNA-seq algorithms can be computationally demanding, especially with large scRNA-seq datasets. Efficient algorithms are needed to scale up with data size and parallelize computation. To improve scalability, algorithms can be designed to run in parallel, distributing the workload across multiple processors. However, not all methods are easily parallelizable, and implementing parallel processing requires careful consideration of data dependencies and synchronization. Also, storing large scRNA-seq datasets, including intermediate results and multiple versions of data processed through different pipelines, requires substantial storage capacity. Efficient access to data is crucial. Methods must be developed to quickly retrieve relevant subsets of data without loading entire datasets into memory.

Integration of multiple data types

Integrating scRNA-seq data with other omics (e.g., proteomics, epigenomics, interactomics) can provide more comprehensive insights but also requires sophisticated computational methods to handle different data types and scales.

Inferring complex trajectory topologies

Current methods often struggle with accurately inferring complex trajectory structures such as tree-like, cyclic, or disconnected topologies. There is a tendency to either overestimate or underestimate the complexity of the underlying topology. Even when dimensionality reduction methods suggest a clear trajectory, many inference methods fail to capture the correct topology, indicating a gap between dimensionality reduction and accurate trajectory inference [12].

Cluster selection and rare cell types

The optimal trajectory, which aligns with the cell connectivity profile, might not be easily derived from the top-ranking clustering algorithms. This difficulty arises due to the variability in cell type abundance. Rare cell types might not be adequately captured in the clustering process if the number of clusters is too low. This under-representation can lead to an inaccurate trajectory topology. Further, the overrepresentation of certain cell types can skew the clustering results, making it challenging to discern the correct trajectory [14].

Validation and Benchmarking

Obtaining ground truth data for benchmarking trajectory inference methods is challenging. Experimental validation of predicted trajectories and cell fates often requires complex and time-consuming experiments. There is a need for robust benchmarking tools and standardized datasets to evaluate and compare the performance of different methods.

Biological interpretation

There are various challenges pertaining to biological interpretation, including feature selection, trajectory uncertainty, and functional validation. Identifying which genes are most informative for trajectory inference is non-trivial. Poor gene (feature) selection can lead to incorrect or ambiguous trajectories. Techniques such as PCA, t-SNE, or UMAP are often used to reduce dimensionality, but the choice of method can affect the inferred trajectory. Further, accurately identifying branching points in differentiation paths and assigning pseudotime (a temporal ordering of cells) can be difficult. Different methods may produce different branching structures and pseudotime estimations. Furthermore, functional validation of inferred trajectories and predicted cell fates through experimental methods (e.g., lineage tracing) is essential but resource-intensive.

5.7 Conclusion

Trajectory inference and cell fate prediction, using scRNA-seq data, allow researchers to map cell progression through different states and predict future cell states based on gene expression profiles. These techniques have broad applications in medical sciences, providing insights into cellular development, disease mechanisms, and personalized medicine. Machine learning plays a crucial role in inference and cell fate prediction. This chapter presented various state-of-the-art methods for trajectory inference and cell fate prediction. In this chapter, we presented the analysis of scRNA-seq data for trajectory inference from aggressive neuroendocrine small cell prostate cancer trans-differentiation using the PAGA software tool. Trajectory inference via PAGA provided a comprehensive view of cell transitions and heterogeneity, revealing distinct and overlapping stages of cancer growth. These insights help identify key genes and pathways for further study, aiding in cancer diagnosis and treatment strategies. Further, some of the challenges and future work directions for trajectory inference and cell fate predictions have been discussed.

References

1. S. Aibar, C.B. González-Blas, T. Moerman, V.A. Huynh-Thu, H. Imrichova, G. Hulselmans, S. Aerts et al., SCENIC: single-cell regulatory network inference and clustering. Nat. Methods **14**(11), 1083–1086 (2017)
2. V. Bergen, M. Lange, S. Peidli, F.A. Wolf, F.J. Theis, Generalizing RNA velocity to transient cell states through dynamical modeling. Nat. Biotechnol. **38**(12), 1408–1414 (2020)
3. J. Cao, M. Spielmann, X. Qiu, X. Huang, D.M. Ibrahim, A.J. Hill, J. Shendure et al., The single-cell transcriptional landscape of mammalian organogenesis. Nature **566**(7745), 496–502 (2019)
4. J.H. Du, T. Chen, M. Gao, J. Wang, Joint trajectory inference for single-cell genomics using deep learning with a mixture prior. bioRxiv, 2020-12 (2020)
5. L. Haghverdi, L.S. Ludwig, Single-cell multi-omics and lineage tracing to dissect cell fate decision-making. Stem Cell Rep. **18**(1), 13–25 (2023)
6. K. Kamimoto, B. Stringa, C.M. Hoffmann, K. Jindal, L. Solnica-Krezel, S.A. Morris, Dissecting cell identity via network inference and in silico gene perturbation. Nature **614**(7949), 742–751 (2023)
7. W. Kyaw, R.C. Chai, W.H. Khoo, L.D. Goldstein, P.I. Croucher, J.M. Murray, T.G. Phan, ENTRAIN: integrating trajectory inference and gene regulatory networks with spatial data to co-localize the receptor-ligand interactions that specify cell fate. Bioinformatics (Oxford, Engl.) **39**(12), btad765 (2023)
8. M. Lange, V. Bergen, M. Klein, M. Setty, B. Reuter, M. Bakhti, F.J. Theis et al., Cell rank for directed single-cell fate mapping. Nat. Methods **19**(2), 159–170 (2022)
9. J. Lee, N. Kim, K.H. Cho, Decoding the principle of cell-fate determination for its reverse control. NPJ Syst. Biol. Appl. **10**, 47 (2024). https://doi.org/10.1038/s41540-024-00372-2
10. Q. Li, scTour: a deep learning architecture for robust inference and accurate prediction of cellular dynamics. Genome Biol. **24**(1), 149 (2023)
11. H. Roux de Bézieux, K. Van den Berge, K. Street, S. Dudoit, Trajectory inference across multiple conditions with condiments. Nat. Commun. **15**(1), 833 (2024)
12. W. Saelens, R. Cannoodt, H. Todorov, Y. Saeys, A comparison of single-cell trajectory inference methods. Nat. Biotechnol. **37**(5), 547–554 (2019)

13. M. Setty, V. Kiselivas, J. Levine, A. Gayoso, L. Mazutis, D. Pe'er, Characterization of cell fate probabilities in single-cell data with Palantir. Nat. Biotechnol. **37**(4), 451–460 (2019)
14. J. Smolander, S. Junttila, L.L. Elo, Cell-connectivity-guided trajectory inference from single-cell data. Bioinformatics (Oxford, Engl.) **39**(9), btad515 (2023)
15. C.J. Soelistyo, G. Vallardi, G. Charras, A.R. Lowe, Learning biophysical determinants of cell fate with deep neural networks. Nat. Mach. Intell. **4**(7), 636–644 (2022)
16. S.V. Stassen, G.G. Yip, K.K. Wong, J.W. Ho, K.K. Tsia, Generalized and scalable trajectory inference in single-cell omics data with VIA. Nat. Commun. **12**(1), 5528 (2021)
17. K. Street, D. Risso, R.B. Fletcher, D. Das, J. Ngai, N. Yosef, S. Dudoit et al., Slingshot: cell lineage and pseudotime inference for single-cell transcriptomics. BMC Genom. **19**, 1–16 (2018)
18. C. Trapnell, Defining cell types and states with single-cell genomics. Genome Res. **25**(10), 1491–1498 (2015)
19. P. Weiler, M. Lange, M. Klein, D. Pe'er, F. Theis, Unified fate mapping in multiview single-cell data. bioRxiv, 2023-07 (2023)
20. F.A. Wolf, P. Angerer, F.J. Theis, SCANPY: large-scale single-cell gene expression data analysis. Genome Biol. **19**, 1–5 (2018)
21. F.A. Wolf, F.K. Hamey, M. Plass, J. Solana, J.S. Dahlin, B. Göttgens, F.J. Theis et al., PAGA: graph abstraction reconciles clustering with trajectory inference through a topology preserving map of single cells. Genome Biol. **20**, 1–9 (2019)
22. Y. Zhang, D. Tran, T. Nguyen, S.M. Dascalu, F.C. Harris Jr., A robust and accurate single-cell data trajectory inference method using ensemble pseudotime. BMC Bioinform. **24**(1), 55 (2023)

Chapter 6
Emerging Topics and Future Directions

6.1 Introduction

Single-cell RNA sequencing (scRNA-seq) is a powerful technology that enables researchers to measure gene expression at the resolution of individual cells. It helps us characterize cellular heterogeneity within complex tissues, identify rare cell populations, and study dynamic processes such as cell differentiation and response to stimuli. Applications of scRNA-seq include cell type identification, differential expression analysis to identify diagnostic and therapeutic markers, trajectory inference to understand developmental processes, dissecting tumor heterogeneity, unraveling immune cell responses, and discovering novel cell states in various biological contexts. Hence, scRNA-seq revolutionizes our understanding of cellular diversity and dynamics, offering insights into fundamental biological processes and disease mechanisms.

Some of the challenges associated with scRNA-seq data analysis include *managing batch effects and technical variability, handling high-dimensional data, accurately identifying cell types and clusters, inferring temporal dynamics and trajectories, cell fate prediction, integrating multi-modal data sources,* addressing *data sparsity and dropout events,* and ensuring *scalability of computational methods* to handle large datasets efficiently. Additionally, interpretation and validation of results, standardization of analysis pipelines, and ethical considerations regarding data privacy and sharing are also important challenges in the field.

Machine Learning and deep learning algorithms plays a crucial role in scRNA-seq analysis by providing robust and scalable computational methods for tasks such as data integration, dimensionality reduction, cell type identification, trajectory inference, and predictive modeling [10]. These algorithms enable the extraction of meaningful patterns from high-dimensional single-cell data, aiding in the identification of cellular heterogeneity, characterization of dynamic processes, and discovery of novel cell states. Machine learning techniques also facilitate the integration of multi-modal omics data and help overcome challenges such as batch effects, data sparsity,

© The Author(s), under exclusive license to Springer Nature Singapore Pte Ltd. 2024   81
K. Raza, *Machine Learning in Single-Cell RNA-seq Data Analysis*,
SpringerBriefs in Computational Intelligence,
https://doi.org/10.1007/978-981-97-6703-8_6

and noise. The purpose of this chapter is to present emerging topics in scRNA-seq analysis and future work directions.

6.2 Emerging Topics in scRNA-seq Analysis

Emerging topics in scRNA-seq analysis include advancements in single-cell multi-omics data integration, spatial transcriptomics, time-series scRNA-seq techniques, development of deep learning architectures such as graph neural networks (GNNs) and autoencoders for various analysis tasks. Transfer learning approaches are gaining attention for leveraging knowledge from related datasets to improve analysis of single-cell data. There's a growing interest in integrating spatial and temporal information into machine learning models for comprehensive understanding of cellular dynamics. Additionally, efforts are being made towards standardization and benchmarking of machine learning methods in single-cell analysis to ensure reproducibility and reliability across studies. Some of the emerging topics in scRNA-seq data analysis are discussed in the following sections (Fig. 6.1).

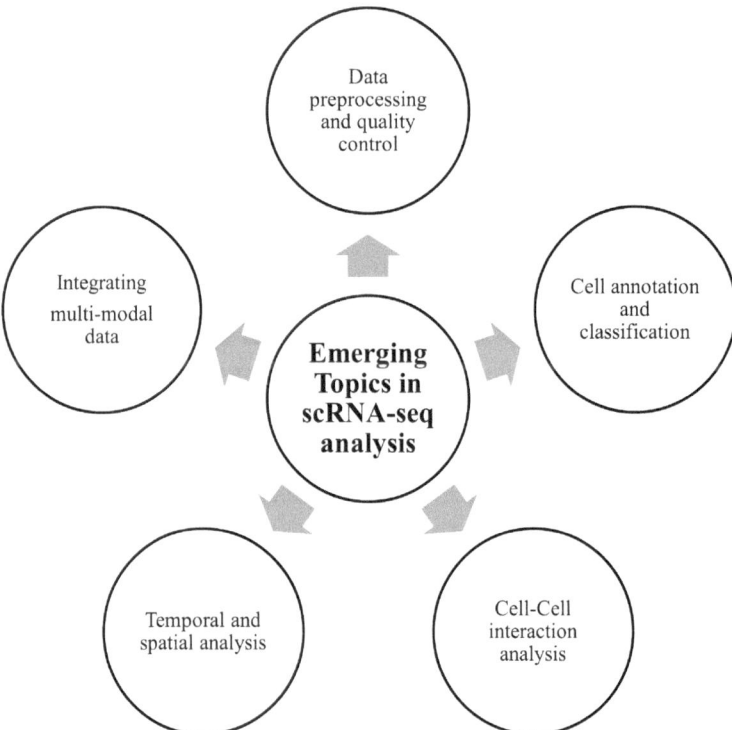

Fig. 6.1 Emerging topics in scRNA-seq analysis

6.2.1 Data Preprocessing and Quality Control

Machine learning-based imputation methods are instrumental in handling dropouts and batch effects in scRNA-seq data. These methods leverage algorithms such as deep learning, probabilistic models, and domain adaptation techniques to effectively address these challenges. For dropout imputation, deep learning approaches like variational autoencoders (VAEs) [4] and generative adversarial networks (GANs) [12] can learn latent representations of the data and generate realistic imputations for missing values. Probabilistic models, such as Bayesian matrix factorization [14], incorporate uncertainty estimates to provide more accurate imputations. In batch effect correction, machine learning methods like domain adaptation techniques [10, 19] and causal GANs [18] can identify and remove batch effects while preserving biological variability. These methods learn batch-specific features and adjust expression profiles to harmonize data across batches.

6.2.2 Cell Annotation and Classification

Cell annotation and classification offer several applications, including identification of distinct cell types and states, aiding in unraveling cellular heterogeneity and lineage relationships. This knowledge is crucial for studying developmental processes, disease mechanisms, and tissue regeneration. Moreover, accurate cell annotation facilitates the discovery of novel biomarkers and therapeutic targets, leading to advancements in personalized medicine and drug development. Numerous methods have been developed for single-cell identity annotation tasks, categorized as *marker-based annotation* and *reference-based annotation* [7]. Marker-based annotation relying on marker genes associated with cell types, and reference-based annotation utilizes a reference dataset for similarity comparisons. Some of the recently developed machine learning methods and tools includes scAnnoX [7], CTISL [11], ICCELF [3], scMMT [17]. However, future improvement in the area is required.

6.2.3 Cell–Cell Interaction Analysis

Cell–cell interaction (CCI) study focuses on understanding the communication and interactions between individual cells. It explores how cells interact with each other through signaling molecules, cell surface receptors, and physical contacts to regulate various physiological processes, such as development, immune response, and tissue homeostasis. scRNA-seq plays a pivotal role in cell–cell interaction studies by providing high-resolution gene expression profiles of individual cells. By analyzing scRNA-seq data, researchers can infer potential interactions between cells based on the expression patterns of signaling molecules, receptors, and ligands. Some of the recently developed CCI methods are SEnSCA [16] for ligand-receptor interactions, MMCCI for multi-model CCI [6], CPPLS-MLP [15] for CCI networks construction and identification of highly variable genes based on scRNA-seq and spatial transcriptomics, and CellCommuNet [9] as a comprehensive database resource facilitating the exploration of cell–cell communication networks.

While scRNA-seq provides insights into cellular heterogeneity, existing methods for CCI analysis often aggregate cell groups, overlooking interactions at the single-cell level. There is a pressing need for analytical techniques that dissect CCI pathways at the single-cell resolution, leveraging the full information content of scRNA-seq data, incorporate spatial information, gene categorization, and integrate multimodal data [13].

6.2.4   Temporal and Spatial Analysis

Temporal and spatial analysis in scRNA-seq refers to the study of gene expression patterns across time or space, respectively, at the single-cell level. These analyses provide insights into dynamic changes in gene expression over time or spatial organization of cells within tissues. Temporal analysis involves tracking changes in gene expression patterns over time. Techniques for temporal analysis in scRNA-seq include pseudotime analysis (inferring the temporal order of cells along a biological process using algorithms like Monocle or Slingshot, already discussed in previous chapter), and time-series analysis (analyzing gene expression profiles across multiple time points to identify dynamically regulated genes and pathways). Spatial analysis involves investigating the spatial organization of cells within tissues to understand their spatial relationships and interactions. Techniques for spatial analysis in scRNA-seq include spatial transcriptomics (mapping gene expression profiles onto tissue sections using methods like Slide-seq or Visium), and image analysis (extracting spatial information from tissue images to correlate cell positions with gene expression patterns). Some of the challenges of temporal and spatial analysis are (i) ensuring accurate alignment of cells across different time points or conditions, (ii) managing technical and biological variability inherent in scRNA-seq data, (iii) achieving high spatial resolution to accurately map gene expression patterns at the single-cell level, (iv) analyzing spatial organization in heterogeneous tissues with diverse cell types and structures.

6.2.5   Integrating Multi-modal Data

Integrating multi-modal data in scRNA-seq analysis involves combining information from different omics modalities, such as gene expression, chromatin accessibility, DNA methylation, or protein expression, to gain a more comprehensive understanding of cellular phenotypes and regulatory mechanisms at the single-cell level.

The benefits of this integration are enhanced resolution, complementary information, and improved interpretability.

In recent years, several single-cell multi-omics data integration tools have emerged to comprehensively characterize cell states and analyze gene regulation. These tools are MOFA (unsupervised integration of multi-omics data) [1], totalVI (a framework for end-to-end analysis using joint probabilistic modeling) [5], GLUE (graph-linked unified embedding for multi-omics single-cell data integration and regulatory inference) [2] and MSDL (multi-use deep learning method) [8].

6.3   Future Directions

Future directions for machine learning-based approaches for scRNA-seq data analysis includes, but not limited to:

Handling batch effects and technical variability: We need to develop robust machine learning methods that can effectively handle batch effects and technical variability in scRNA-seq data, especially in large-scale studies involving multiple experimental batches or datasets.

Cell type annotation and classification: Some of the limitations existing approaches for cell type annotation and classification are operational efficiency, optimizing clustering efficiency. These limitations can be addressed by leveraging anchor graph technology to expedite the learning process and reduce runtime (). Development of more accurate and robust algorithms for cell type annotation, leveraging advancements in machine learning, deep learning, and data integration techniques, are needed. These algorithms should be able to handle complex biological heterogeneity and noisy single-cell data more effectively.

Single-cell trajectory inference: Develop improved methods for cell trajectory inference. This could involve the integration of temporal information and the incorporation of prior knowledge about cell fate decisions and lineage relationships.

Multimodal data integration: Combining data from different omics modalities requires robust computational methods for data alignment, normalization, and integration, especially considering differences in data sparsity and noise levels. There is a need to develop more advanced methods that can effectively integrate scRNA-seq data with other omics modalities, including single-cell ATAC-seq, single-cell DNA methylation, or spatial transcriptomics data. This would enable a more comprehensive understanding of cellular states and regulatory mechanisms.

Spatial transcriptomics: Develop advanced machine learning approaches for analyzing spatial transcriptomics data, which provide information about the spatial organization of cells within tissues. This could involve techniques for spatial clustering, spatial gene expression prediction, and spatial trajectory inference.

Deep learning architectures: Exploration of novel deep learning architectures for scRNA-seq data analysis. This could involve the development of deep generative models for imputation and denoising, as well as graph neural networks (GNNs) for modeling cellular interactions and regulatory networks.

Interpretability and explainability: Enhancement of interpretability and explainability of machine learning models are needed. This could involve the development of methods for identifying biologically meaningful features, interpreting model predictions, and understanding the underlying regulatory mechanisms.

Transfer learning and domain adaptation: Explore transfer learning and domain adaptation techniques to leverage information from existing scRNA-seq datasets for analyzing new datasets, especially in cases where labeled data is limited or unavailable.

Standardization and benchmarking of machine learning methods: Standardization and benchmarking of machine learning methods are essential for ensuring the reliability, reproducibility, transparency, and comparability of computational tools and algorithms developed for analyzing scRNA-seq data. Standardization involves establishing standardized procedures for data preprocessing steps (quality control, normalization, feature selection, and batch correction), implementing algorithms and computational methods (following best practices and guidelines), data formats and representations used for storing and sharing scRNA-seq data. Benchmarking involves systematically evaluating the performance of machine learning methods using standardized datasets and evaluation metrics. This allows researchers to objectively compare the performance of different methods in terms of accuracy, efficiency, scalability, and robustness. It would make informed decisions when selecting tools for analyzing scRNA-seq data.

Building robust and scalable machine learning pipelines: It involves designing a systematic workflow that can handle the complexity and scale of scRNA-seq data while ensuring reliability, reproducibility, and efficiency. Choose appropriate machine learning models based on the specific analysis task, such as clustering, classification, dimensionality reduction, trajectory inference, or differential expression analysis, etc. Also consider the characteristics of scRNA-seq data, such as sparsity, high dimensionality, and noise, when selecting models. Design the pipeline as a series of modular components (e.g., data preprocessing, feature engineering, model training, evaluation) that can be easily combined and adapted for different analysis tasks. Ensure that the pipeline can handle large-scale scRNA-seq datasets efficiently by optimizing computational resources, parallelizing computations, and implementing distributed computing techniques if necessary. Optimize model hyperparameters using techniques such as grid search, random search, or Bayesian optimization to improve model performance. By following best practices and incorporating these key elements, researchers can develop pipelines that effectively analyze scRNA-seq data and derive meaningful biological insights.

Interactive and user-friendly tools: Design interactive and user-friendly software workbench for integrative analysis of scRNA-seq data, which enable biologists and researchers with limited computational expertise to perform sophisticated analyses and gain insights from their data.

6.4   Conclusion

In conclusion, the advent of scRNA-seq has propelled the field of cellular biology by enabling unprecedented insights into cellular heterogeneity and dynamics. Machine learning-based approaches play a pivotal role in harnessing the full potential of scRNA-seq data, offering robust and scalable solutions to address key challenges in data analysis. Through this chapter, we have explored emerging topics and future directions in machine learning for scRNA-seq analysis, covering a wide range of applications such as data preprocessing, cell type annotation, cell–cell interaction analysis, and integration of multi-modal data. We have highlighted the importance of standardization and benchmarking of machine learning methods to ensure reproducibility and reliability across studies. Furthermore, the development of robust and scalable machine learning pipelines is essential to facilitate the analysis of increasingly large and complex scRNA-seq datasets. By embracing these advancements and fostering interdisciplinary collaborations, we can unlock new frontiers in disease understanding, and precision medicine. Moving forward, continued innovation and collaboration in the field of machine learning and scRNA-seq analysis holds promise for unraveling the complexities of cellular systems and advancing biomedical research.

References

1. A.S.E. Cuomo, T. Heinen, D. Vagiaki, D. Horta, J.C. Marioni, O. Stegle, Cell RegMap: a statistical framework for mapping context-specific regulatory variants using scRNA-seq. Mol. Syst. Biol. **18**(8), e10663 (2022)
2. Z.J. Cao, G. Gao, Multi-omics single-cell data integration and regulatory inference with graph-linked embedding. Nat. Biotechnol. **40**(10), 1458–1466 (2022)
3. S. Cui, S. Nassiri, I. Zakeri, Imbalance and composition correction ensemble learning framework (ICCELF): a novel framework for automated scRNA-seq cell type annotation. bioRxiv, 2024-04 (2024)
4. R. Danino, I. Nachman, R. Sharan, Batch correction of single-cell sequencing data via an autoencoder architecture. Bioinform. Adv. **4**(1), vbad186 (2024)
5. A. Gayoso, Z. Steier, R. Lopez, J. Regier, K.L. Nazor, A. Streets, N. Yosef, Joint probabilistic modeling of single-cell multi-omic data with totalVI. Nat. Methods **18**(3), 272–282 (2021)
6. L. Hockey,, O. Mulay, Z. Xiong, K. Khosrotehrani, C.M. Nefgzer, Q. Nguyen, MMCCI: multi-modal cell–cell interaction integrative analysis of single cell and spatial transcriptomics data. bioRxiv, 2024-02 (2024)
7. X. Huang, R. Liu, S. Yang, X. Chen, H. Li, ScAnnoX: an R package integrating multiple public tools for single-cell annotation. PeerJ **12**, e17184 (2024)

8. J. Lakkis, A. Schroeder, K. Su, M.Y.Y. Lee, A.C. Bashore, M.P. Reilly, M. Li, A multi-use deep learning method for CITE-seq and single-cell RNA-seq data integration with cell surface protein prediction and imputation. Nat. Mach. Intell. **4**(11), 940–952 (2022)

9. Q. Ma, Q. Li, X. Zheng, J. Pan, Cell CommuNet: an atlas of cell–cell communication networks from single-cell RNA sequencing of human and mouse tissues in normal and disease states. Nucleic Acids Res. **52**(D1), D597–D606 (2024)

10. I. Rivero-Garcia, M. Torres, F. Sánchez-Cabo, Deep generative models in single-cell omics. Comput. Biol. Med. **176**, 108561 (2024)

11. X. Wang, Z. Chai, S. Li, Y. Liu, C. Li, Y. Jiang, Q. Liu, CTISL: a dynamic stacking multi-class classification approach for identifying cell types from single-cell RNA-seq data. Bioinformatics **40**(2), btae063 (2024)

12. Y. Wang, T. Liu, H. Zhao, ResPAN: a powerful batch correction model for scRNA-seq data through residual adversarial networks. Bioinformatics **38**(16), 3942–3949 (2022)

13. A.J. Wilk, A.K. Shalek, S. Holmes, C.A. Blish, Comparative analysis of cell–cell communication at single-cell resolution. Nat. Biotechnol. **42**(3), 470–483 (2024)

14. Y. Xu, Z. Zhang, L. You, J. Liu, Z. Fan, X. Zhou, ScIGANs: single-cell RNA-seq imputation using generative adversarial networks. Nucleic Acids Res. **48**(15), e85 (2020)

15. T. Zhang, Z. Wu, L. Li, J. Ren, Z. Zhang, G. Wang, CPPLS-MLP: a method for constructing cell–cell communication networks and identifying related highly variable genes based on single-cell sequencing and spatial transcriptomics data. Brief. Bioinform. **25**(3), bbae198 (2024)

16. L. Zhou, X. Wang, L. Peng, M. Chen, H. Wen, SEnSCA: identifying possible ligand-receptor interactions and its application in cell–cell communication inference. J. Cell. Mol. Med. **28**(9), e18372 (2024)

17. S. Zhou, Y. Li, W. Wu, L. Li, ScMMT: a multi-use deep learning approach for cell annotation, protein prediction and embedding in single-cell RNA-seq data. Brief. Bioinform. **25**(2), bbad523 (2024)

18. Y. Zinati, A. Takiddeen, A. Emad, GRouNdGAN: GRN-guided simulation of single-cell RNA-seq data using causal generative adversarial networks. Nat. Commun. **15**(1), 4055 (2024)

19. B. Zou, T. Zhang, R. Zhou, X. Jiang, H. Yang, X. Jin, Y. Bai, DeepMNN: deep learning-based single-cell RNA sequencing data batch correction using mutual nearest neighbors. Front. Genet. **12**, 708981 (2021)